T0205407

Studies in Systems, Decision and Control

Volume 425

Series Editor

Janusz Kacprzyk, Systems Research Institute, Polish Academy of Sciences,
Warsaw, Poland

The series "Studies in Systems, Decision and Control" (SSDC) covers both new developments and advances, as well as the state of the art, in the various areas of broadly perceived systems, decision making and control–quickly, up to date and with a high quality. The intent is to cover the theory, applications, and perspectives on the state of the art and future developments relevant to systems, decision making, control, complex processes and related areas, as embedded in the fields of engineering, computer science, physics, economics, social and life sciences, as well as the paradigms and methodologies behind them. The series contains monographs, textbooks, lecture notes and edited volumes in systems, decision making and control spanning the areas of Cyber-Physical Systems, Autonomous Systems, Sensor Networks, Control Systems, Energy Systems, Automotive Systems, Biological Systems, Vehicular Networking and Connected Vehicles, Aerospace Systems, Automation, Manufacturing, Smart Grids, Nonlinear Systems, Power Systems, Robotics, Social Systems, Economic Systems and other. Of particular value to both the contributors and the readership are the short publication timeframe and the worldwide distribution and exposure which enable both a wide and rapid dissemination of research output.

Indexed by SCOPUS, DBLP, WTI Frankfurt eG, zbMATH, SCImago.

All books published in the series are submitted for consideration in Web of Science.

More information about this series at https://link.springer.com/bookseries/13304

Igor Ruban · Andriy Kovalenko · Vitaly Levashenko
Editors

Advances in Self-healing Systems Monitoring and Data Processing

 Springer

Editors
Igor Ruban
Kharkiv National University of Radio
Electronics
Kharkiv, Ukraine

Andriy Kovalenko
Kharkiv National University of Radio
Electronics
Kharkiv, Ukraine

Vitaly Levashenko
Faculty of Management Science
and Informatics
University of Žilina
Žilina, Slovakia

ISSN 2198-4182 ISSN 2198-4190 (electronic)
Studies in Systems, Decision and Control
ISBN 978-3-030-96548-8 ISBN 978-3-030-96546-4 (eBook)
https://doi.org/10.1007/978-3-030-96546-4

This Springer imprint is published by the registered company Springer Nature Switzerland AG
The registered company address is: Gewerbestrasse 11, 6330 Cham, Switzerland

Contents

Self-healing Systems Monitoring

Igor Ruban, Vitalii Martovytskyy, and Olesia Barkovska

Abstract Monitoring the condition of Self-healing Systems is a compulsory system component. The authors proposed an approach to identifying anomalies in ShS operation based on machine learning technology. The proposed architecture of the monitoring system using autonomous software agents. The architecture provides for the dynamic development of a hierarchical structure, the node of which can be any entity that is determined by the data source or sensor. For interaction among all agents, it is proposed to use a group of intelligent query agents whose purpose is to coordinate information gathering agents, restructure the received information and implement protocols and messaging mechanisms among all agents of the model. In the context of ShS monitoring, there may exist the metrics of grids, clusters, computational nodes and tasks, and so on. Based on this approach, a methodology for monitoring the system condition is proposed. The proposed methodology determines the conditions and the procedure for assessing the ShS condition using the developed multi-agent monitoring system.

Keywords Self-healing systems · Condition monitoring · Intelligent agent · Anomaly · Multi-agent system

1 Principles of a Self-healing Systems Monitoring System

Nowadays there are a lot of monitoring systems [1–12]. For example, Nagios system [1] is designed to monitor computer systems and networks. It scans target nodes and services and informs an administrator if one of the services stops operating or resumes it again. The drawbacks of this system are poor scalability, a large interval among parameter measurements, averaged data, and no tools of automated expert data

I. Ruban · V. Martovytskyy (✉) · O. Barkovska
Kharkiv National University of Radio Electronics, 14 Nauki ave, Kharkiv 61166, Ukraine

I. Ruban
e-mail: ihor.ruban@nure.ua

O. Barkovska
e-mail: olesia.barkovska@nure.ua

analysis. Zabbix system [2] is designed to monitor and track the statuses of various services of a computer network, servers, and network equipment. This monitoring system is stable and reliable and has a steady rate of development. But there also exist poor scalability, low fault tolerance, and no data analysis tools. In addition, there are problems with the integration of this software product.

The system of monitoring proposed by the author of article [3], based on the multi-agent approach, handles the problems of scaling and fault-tolerance. The author proposes to create agents that are scattered across computational nodes and collect data on system performance. The disadvantage of such monitoring is the incompleteness of the analysis since it deals with computational nodes only. A dynamically reconfigurable distributed modular monitoring system is proposed in article [4]. Along with the agent approach, the author suggests using statistical collection of data from the systems where an agent cannot be created for some reasons. The drawback of this system is the unstandardized data obtained from various sources, which results in complicated data mining and system integration. In article [5], the authors proposed the instrumental complex of meta-monitoring distributed computing environments. But for data mining, the expert system is used, developed using CLIPS shell. There is no self-learning mechanism in this expert system, due to this the system cannot detect new anomalies. The author of article [6], together with Nagios, uses neural networks to detect anomalies, which solves the problem of learning, but does not handle other drawbacks of this system.

So, the analysis of the above systems showed that today there is a need to develop a subsystem for monitoring the ShS condition, which, based on data on the system operation could assess the integrity of the ShS computational process. This monitoring subsystem must have a simple process of integration with ShS and meet the principles of completeness, scalability, fault tolerance, standardization of processed data, and have a self-learning mechanism.

2   Model of Self-healing Systems Monitoring Subsystem Operation Based on a Client–Server Architecture

2.1   Peculiarities of a Client–Server Model Implementation

Consider a general client–server model in which the client initiates data exchange by sending a request to the server. The server processes the request and, if necessary, sends a response to the client, as is shown in Fig. 1.

Fig. 1   A client–server model

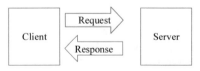

The client–server model is a technology of interaction among computers in a network, in which each computer has its own purpose and fulfils its own functions. Some computers in the network have information and computing resources and control administer them, while others can access these resources through services.

The client–server architecture combines system usability, the ability to apply user capacity as well as the high performance that the server can provide during data sharing operations. In this case, the type of operating system and equipment used are not key factors. The main requirement is to use the same protocols and applications for data exchange [13].

The client–server interaction can be either synchronous when the client waits for the moment when the server finishes processing the request, or asynchronous when the client sends a request to the server and continues functioning without the server's response. The client–server model can be used as a basis for describing various interactions. To analyse the structure of ShS operation, the interaction of software components that form a distributed system is important.

To ensure efficient operation, it is necessary to monitor actively the network as well as the network clients' actions that search problems caused by overloaded or failed servers, other devices or network connections, unauthorized changes to databases, system parameters, and so on.

In distributed computer systems, where the components interact within the client–server model, it is necessary to monitor the clients' actions to control the integrity of data, maintain statistical records of errors, single out popular and irrelevant system functions, and so on.

It is necessary to carry out the control of applications that specifies errors and application system messages as well as information about all the actions of users. In addition, all suspicious events during the operation of the system must be recorded.

User's rights to access ShS resources must be also monitored. This will enable identifying attempts to access resources with deliberately insufficient rights [14].

There are a large number of monitoring methods and tools that enable the control of various components of the system that is based on a client–server model, however, some general elements of any monitoring can be singled out within the client–server interaction model.

The first stage of monitoring is to test the equipment availability.

The second stage is to the health check of critical services running on the network. After that, the parameters specific to the services for a particular system are checked.

The third stage is to check the integrity of databases. In addition, a certain set of events is also monitored, among them are OS events, system registry events, OS kernel-layer events, session-layer events, events at the layer of external interaction, events at the layer of direct access to application logic, DBMS-layer events.

Thus, the model of monitoring ShS clients is shown in Fig. 2.

Consider a typical application, which, according to modern concepts, can be divided into the following logical layers, which are shown in Fig. 3: user interface—data presentation layer, application logic—business logic layer, and data access—data layer that works with a database.

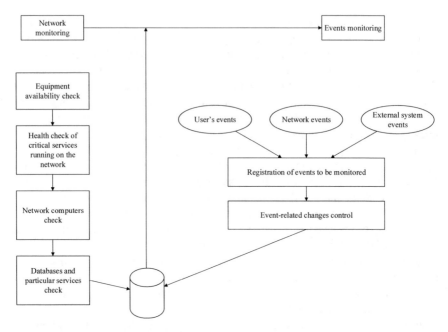

Fig. 2 Model of a client–server monitoring

Fig. 3 Typical application layers

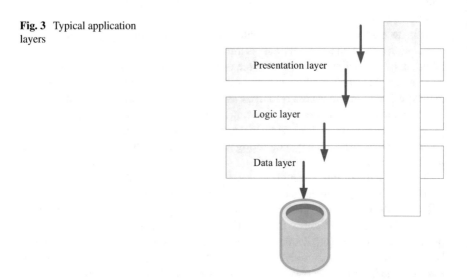

Three layers (in the context of programming) are storage, processing, and storage of information. The idea is not to confuse these layers. The three-layer approach is good programming practice that can be used to develop almost any application. In distributed applications, these three layers are physically located on different

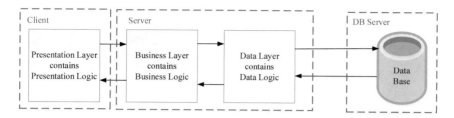

Fig. 4 Two-stage architecture

computers, and it is also possible to have several interchangeable variants of each layer.

Since in practice, different users of the system are usually interested in accessing the same data, the functions of such a system can be most easily distributed among several computers by dividing the logical layers of the application among one program server responsible for accessing data and several client computers that implement the interface. The logic of the program can be assigned to the server, clients, or shared among them, as is shown in Fig. 4.

The architecture of applications based on this principle is called a client–server or two-stage. Such systems are not often classified as distributed in practice, but formally they can be considered the simplest distributed systems.

Also, systems with such an architecture are more secure, since they can use their own data transfer protocols, unknown to intruders. Therefore, many large companies that are not located in a single place and use the global Internet to connect business sectors choose this architecture to build client/server systems.

The above-stated distributed architecture of a computer system with databases can be advanced in a three-layer model with an application server. In this case, in addition to the database server and client computers, another independent component is singled out in the system, the so-called application server, which is placed between the client subsystems and the database server, as is shown in Fig. 5.

The user's request is sequentially processed in such systems by the system client part, the application logic server, and the database server. But as a rule, systems with a more complex architecture are considered a distributed system. At the same time, the three-layer model has found a new impetus for development and begun actively evolving to a multi-layer one, being gradually complicated by adding new various intermediate services. Moreover, as a result of symbiosis, such a system has not only features and solutions characteristic of web technologies, but also completely new qualities and properties [15].

Consider a direct client-to-client communications network as a simple example. A peer-to-peer network is a single-layer network where absolutely all network nodes perform the same functions or can automatically change their set of functions depending on the external environment [16]. Direct exchange networks are organized in a more complex way, as is shown in Fig. 6.

Let us further detail the forming of parameters to monitor infrastructure and applications in ShS.

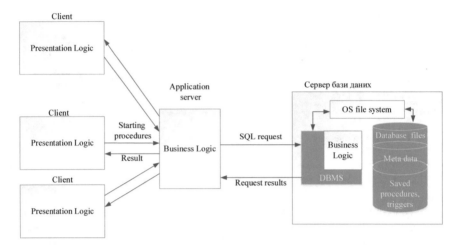

Fig. 5 Tree-stage architecture

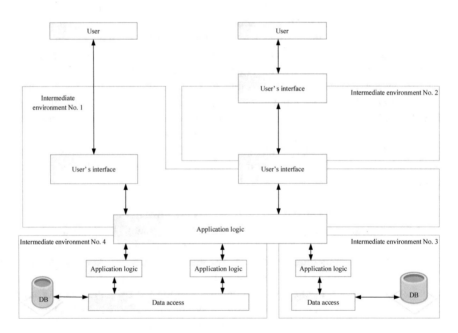

Fig. 6 System of direct data exchange between clients

2.2 Forming Parameters to Monitor Infrastructure and Applications in Self-healing Systems

Understanding the state of the system is essential to enable the reliability and stability of applications and services. Information on system health and capacity helps software engineers deal timely with problems that arise, and also allows them to make changes in configuration changes confidently.

Today, there are many strategies and tools for monitoring ShS components, collecting critical data, and responding to errors and changing conditions in various environments. But as programming methods and infrastructure projects evolve, monitoring must adapt to meet new challenges.

The authors of article [17] single out four monitoring signals that are the most significant factors that must be controlled within the manufacturing system. These monitoring parameters are:

- delay;
- traffic;
- error rate;
- saturation.

Delay is the time necessary to complete an action, and its measurement is specified by the component. Delay indicates how long it will take to complete a specific task or action. Measuring the delay of various components enables building a holistic model of system various characteristics. This can help one find bottlenecks, understand which resources need more time, and pay attention to system slowdowns on time. The authors emphasize in [4] that while calculating delays it is important to distinguish between successful and unsuccessful requests since they can corrupt average service values.

Traffic shows how "busy" components and systems are. This is the load or demand of services that can help understand how much work the system is currently doing. Consistently high or low traffic may indicate that the service might require more resources or that there is an error in the system that affects traffic routing [18, 19]. However, in most cases, traffic data are most useful for understanding problems with other signals. For example, if the delay exceeds an acceptable level, this time interval can be compared to traffic discontinuity. This metric shows the maximum volume of traffic can be processed by the system as well as service failures at various stages.

To have a general picture of the state of components and their response to requests, errors must be monitored. Some applications or services reflect bugs in their built-in interfaces, but additional work may be required to collect data from other programs. Separating different types of errors can enable identifying problems that affect the system services more accurately as well as setting up a flexible notification system—the system can report certain types of errors immediately, and ignore the rest until they exceed a certain threshold.

Saturation enables determining a resource amount is being used. Percentages are often used to determine the saturation of resources that have a clear total capacity, but resources with a less well-defined maximum may require other measurement methods. Saturation data provides information about the ShS resources on which the performance of a service or program depends. Since a service provided by one component can be consumed by another service, saturation is one of the most significant metrics for identifying bandwidth problems. Thus, issues associated to saturation and delays in a single layer can reflect significant traffic discontinuity or errors in the lower layer.

The basic metrics to monitor are the local host parameters of the ShS infrastructure. In modern software engineering, considerable effort is made to abstract the physical components and details of the low-level operating system, but the operation of each service is based on the basic hardware and operating system. Therefore, monitoring basic host resources is the first step towards creating a complete picture of the system state. When selecting host parameters, its specific resources must be taken into consideration. This includes the host hardware as well as the basic components of the operating system such as processes and file descriptors. Considering each component in terms of the four golden signals, it can be seen that some signals are obvious while others are more difficult to assess.

In articles [20–22] suggests a lot of ways to obtain basic Linux parameters. His method of performance analysis is called USE. The USE method and four golden signals are identical to a great extent, so some of his suggestions can be used to figure out what data to collect from server components.

CPU parameters to monitor:

- delay: average delay or maximum delay in the CPU scheduler;
- traffic: CPU load;
- errors: errors of a specific processor, faulty processors;
- saturation: the run queue length.

To monitor memory, signals can be as follows:

- delay: no recommendation—it is difficult to find a good measurement method;
- traffic: the amount of used memory;
- errors: errors in memory;
- saturation: OOM events, swap space use.

To monitor the status of storage devices, the following parameters can be singled out:

- delay: average delay for read and record operations;
- traffic: reading and recording I/O levels;
- errors: system error, disk errors;
- saturation: I/O queue depth.

The network parameters to monitor are:

- delay: network driver queue;
- traffic: input and output bytes or packets per second;

- errors: network device errors, dropped packets;
- saturation: dropped packets, segment retransmissions, congestion.

Along with the parameters of physical resources, it is also necessary to collect indicators of the operating system, which have constraints. Some examples that fall into this category are files and number of threads. These are not physical resources, but constraints set by the operating system to prevent processes over-branching. Most of them can be configured using commands such as *ulimit*. Monitoring changes in the use of these resources can help identify potentially harmful changes in software configurations.

The next level deals with applications and services running on servers. These programs use separate server components that were previously considered as resources. The parameters at this level help assess the state of the host applications and services.

Previous parameters detail the capabilities and performance of individual components and the operating system, while parameters at this level describe the state of software operation. They also show what resources applications depend on and how well they manage constraints.

Monitoring parameters at this level cannot follow the approach that was used at the previous level. The parameters will largely depend on the characteristics of the applications, the configuration and the operational loads of computers. The results will depend on the host requirements.

Application that Apps that serve clients are monitored by the following metrics:

- delay: execution time of requests;
- traffic: the number of requests per second;
- errors: program errors that occur when processing client requests or accessing resources;
- saturation: the percentage or amount of resources that are currently in use.

Dependencies are one of the most important metrics. They are often better expressed by the saturation marks of the individual components. For example, the use of program memory, available connections, a number of files or a number of active operational processes can help assess the impact of the change in the context of a physical server [23].

Recommendations as for monitoring distributed systems presented in book [23] are mainly developed for distributed, therefore they assume a client–server architecture. For applications that do not a client–server architecture, the same parameters are still important, but signal *"traffic"* may need to be slightly reconsidered. This parameter mainly reflects the level of application busy state, so a metric that adequately reflects the application busy state must be determined. The specifics will depend on what a particular program is doing, traffic at this level is usually determined by the number of operations or data that are processed per second.

Most services, especially in a manufacturing environment, can involve several servers to improve performance and availability. This additional layer adds another layer to the monitoring system. Distributed computing and redundant systems can make the infrastructure more flexible, but network-based coordination is more brittle than inter-host communication. Monitoring can help facilitate some difficulties in using a less reliable communication channel.

In addition to the network itself, for distributed servers, the status and performance of the group are more significant than the metrics that are monitored on individual hosts. Services are closely related to the computer on which they are running, if they are limited to one host; however, redundant services running on multiple hosts are assigned to the resources of these hosts and are not directly dependent on a single computer.

At this level the parameters are much similar to the parameters of the previous level, but they additionally consider the interrelation between group members:

- delay: the time the pool needs to respond to requests, the time for communication or synchronization with peer servers;
- traffic: the number of requests processed by the pool per second;
- errors: program errors while processing client requests, accessing resources, or synchronizing with peer servers;
- saturation: the number of currently used resources, the number of running servers, the number of servers available.

The parameters of this level are definitely similar to the parameters of services within the same host. Delay becomes more complex since its monitoring can require communication between multiple nodes. Traffic is no longer limited to the capabilities of a single server, but integrates the capabilities of the group and the efficiency of the routing algorithm. There are additional types of errors related to network connection or host error. Finally, saturation covers the pooled resources of all hosts, the network interconnection between hosts, and the capability to synchronize access to the shared resources that each computer needs.

Some important parameters that need to be monitored are outside ShS that is controlled. These are external dependencies, including those associated with the hosting provider, and any services that ShS programs are assigned to. These programs are resources that ShS cannot directly control, but that can endanger the ShS operation. Parameters to monitor external dependencies include:

- delay: the time necessary to receive a response from the service or to provide new resources from the provider;
- traffic: the number of operations transmitted by the external service, the number of requests received in the external API;
- errors: error ratio for service requests;
- saturation: the amount of resources used that depend on the account, such as instances, API requests, and so on.

These parameters can enable identifying dependency problems, warn about resource depletion, and help control costs. In less flexible situations, the parameters can warn an operator about problems and the need to fix them.

High-level metrics monitor requests through the system in the context of an external component that users interact with. It could be a load balancer or other routing mechanism that is responsible for receiving and synchronizing requests to a service. Since this is the first point of interaction with the system, collecting parameters at this level gives a rough estimate of the overall user experience and enables monitoring user actions. This, in turn, enables identifying a user based on his behaviour.

Parameters to monitor at this level are similar to the above parameters. The basic difference is the scale and significance of data obtained:

- delay: time to complete user requests;
- traffic: the number of user requests per second;
- errors: errors while processing client requests or accessing resources;
- saturation: the percentage or amount of resources used.

Values that fall outside the acceptable ranges for these parameters are likely to indicate direct user impact. Delay that does not meet client's or internal SLAs and traffic that shows a significant jump increase the error rate. The exact parameters at this level can tell a lot about the availability, performance, and reliability of a distributed system API.

2.3 The Structure of the Monitoring Model of the Self-healing Systems Operation

Cluster computing systems are easy to build as it is possible to flexibly adjust the required system performance by connecting regular servers to the cluster using special hardware and software interfaces until a supercomputer of the required power is configured [24]. Clustering enables manipulating a group of servers as one system, and simplifies management and increases the reliability of the system as a whole [25].

A significant feature of clusters is their capability to provide each server with access to any block of both RAM and disk memory. This problem is successfully solved, for example, by combining SMP-architecture systems based on autonomous servers to organize a common field of RAM and using RAID disk systems for external memory [25].

Clusters are usually created by using either simple single-processor personal computers or dual-processor or four-processor SMP servers. At the same time, there are no constraints as for the composition and architecture of the nodes. Each node can operate being controlled by own operating system. The most commonly used are standard operating systems such as Linux, FreeBSD, Solaris, Unix, Windows. In cases where the cluster nodes are heterogeneous, heterogeneous clusters are concerned.

While clustering, two approaches can be singled out.

The first approach is used when creating small cluster systems. Fully functional computers are clustered, continuing to work as independent units. For example, classroom computers or lab workstations.

The second approach is used in cases when a powerful computing resource is purposefully created. Then the system units of computers are compactly placed in special racks, and one or more fully functional computers, called host computers, are assigned to control and run the system.

Having analysed cluster solutions for creating supercomputers from leading companies (IBM, Intel, HP), the following elements were singled out:

- computing nodes;
- high-speed network fabric;
- auxiliary networks (control/monitoring);
- data storage system;
- auxiliary servers (access nodes/compilation nodes/monitoring nodes);
- infrastructure (uninterruptible power supply and cooling systems).

Figure 7 shows the example of ShS structure built by cluster architecture.

Having analysed the ShS architecture (Fig. 7), it can be concluded that the monitoring subsystem must:

- have to be able to integrate with local monitoring tools of computing nodes;
- provide application program interfaces based on open standards for embedding into other software systems;
- include tools of data standardization obtained from various sources;
- have tools for collecting and analysing data on the operation of equipment at every functional element of the cluster;

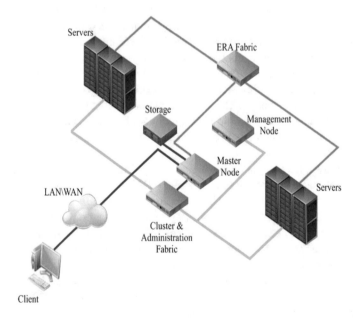

Fig. 7 Typical structure of a distributed computer system

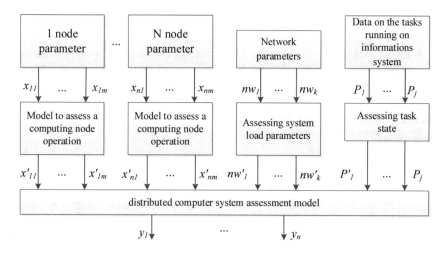

Fig. 8 Structure of the model for monitoring the ShS operation of a cluster architecture to detect anomalies

- provide tools for collecting and analysing data on the distribution of tasks for each user applications;
- monitor the state of a computer network for use in calculations;
- provide tools for automated expert analysis of the data of monitoring and generating control actions.

Taking into account the conditions set, the general structure of the monitoring subsystem has been developed, is shown in Fig. 8.

The parameters of monitoring systems general structure:

- x is the controlled parameter of a computing node;
- nw is the controlled parameter of a computer network;
- P—parameter of the task running on a computer system;
- x'—the assessed parameter of a computing node;
- nw'—the assessed parameter of a computer network;
- P'—parameter for assessing the task running on the computer system;
- y_1—parameter for assessing the state of the computer system.

To describe the monitoring subsystem, the following tuple is proposed:

$$SM = \{\{Ag\}, \{S\}, STK\}, \tag{1}$$

where $\{Ag\}$ is a set of monitoring agents assigned to collect and primary analyse data on key parameters that describe the process of a computing cluster operation: characteristics of running tasks, information about users, network connections and their characteristics, configuration parameters of the kernel modules of every node operating system; $\{S\}$ is a set of system states that is formed based of data obtained by monitoring agents; STK is a system state classifier.

While describing the cluster state, the following can be singled out: the current system state $S_t \in \{S\}$ that is formed based on data obtained by monitoring agents $Ag_j \in \{Ag\}$ in the real time mode and a set of system normal patterns $S_{norm} \subseteq \{S\}$ based on statistical data obtained in the course of system learning when all system states are considered conditionally safe.

Each system state element is described by the following tuple:

$$S = \{\{P\}, \{X\}, \{NW\}\}, \tag{2}$$

where $\{P\}$ is a set of tasks running on the cluster, $\{X\}$ is a set of computing nodes states, $\{NW\}$ is a set of network connections parameters.

In its turn, each task $P_i \in \{P\}$ can be presented as a set of parameters:

$$P_i = \{UID, TID, \{NODE\}, C\}, \tag{3}$$

where UID is a user identifier that has run the task, TID is a task identifier, $\{NODE\}$ is a list of nodes that participate in task processing, C is task execution time.

Each of computing node state of $X_i \in \{X\}$ cluster can be presented as a tuple:

$$X_i = \{\{UIDOS\}, \{PIS\}, \{NOS\}\}, \tag{4}$$

where $\{UIDOS\}$ is a set of users in the node operating system (OS), $\{NOS\}$ is a set of node network connections, $\{PID\}$ is a set of node OS processes.

$PID_i \in \{PID\}$ can be described as the parameters [26]:

$$PID_i = \{UIDOS_i, ID, \{API\}, \{CNP\}\}, \tag{5}$$

where $UIDOS_i \in \{UIDOS\}$ is a user identifier that has run the process (this identifier determines the process rights while performing API functions of the operating system), ID is a process identifier in the system, $\{API\}$ is a set of API functions of the operating system that were called by the application, including the parameters of these calls, $\{CNP\}$ is a set of network connections initiated a specific application.

Data obtained and grouped while monitoring are further fed to the ensemble of STK classifiers, which classifies the system states.

The operation of the classifiers ensemble can be described as follows.

Let there be a set of class marks $Y = \{y_1, y_2,...,y_N\}$ and a set of features describing the state of the system (2). Then a classifier is the following mapping:

$$C : S \rightarrow [0, 1]^N, \tag{6}$$

where $C(s)$ is a dimension vector N whose g-th component determines the degree of the s class membership y_h, $h = 1, ..., N$. In systems based on the combination of q classifiers, the outputs of separate classifiers are combined to obtain the ultimate classifier decision:

$$C(s) = F\big(C_1(s), \ldots, C_q(s)\big), \tag{7}$$

where F is a join operator. The output of each separate classifier for the data object s is the following N-dimensional vector:

$$C(s) = F\big(C_1(s), \ldots, C_q(s)\big). \tag{8}$$

The output of the entire combination of classifiers is the following N-dimensional vector:

$$C(s) = [\varphi_1(s), \ldots, \varphi_N(s)]. \tag{9}$$

If it is necessary to define a single class mark for the object s, then the class y_s corresponds to the maximum value of the degrees of membership:

$$c_{i,s}(s) \geq c_{i,j}(s) \; \forall \, j = 1, \ldots, N \text{---for separate classifiers;} \tag{10}$$

$$\varphi_s(s) \geq \varphi_j(s) \; \forall \, j = 1, \ldots, N \text{---for the entire classifier.} \tag{11}$$

There are different operators that enable combining the outputs of separate classifiers of the ensemble. They include the operators of maximum, minimum, product operator, average operator, majority vote operator and so on.

In this case, the principle of a meta classifier is used as the join operator, and an artificial neural network acts as a meta classifier.

The operation of a multilayer artificial neural network is described by the following tuple:

$$\big(Net_{ij}, Out_{ij}, In_{ijk}, X_k\big), \tag{12}$$

where

$$Net_{ij} = \sum_k w_{ijk} In_{ijk}, \tag{13}$$

$$Out_{ij} = f\big(Net_{ij} - \theta_{ij}\big), \tag{14}$$

$$In_{ijk} = Out_{(i-1)k}, \tag{15}$$

$$In_{0jk} = x_k, \tag{16}$$

where x is a set of input values of a neural network (N-dimensional vector (8)); *In* is a set of input values of a neuron; *Out* is a set of output values of a neuron; i is the number of the neural network layer; j is a the number of a neuron in the neural network layer; k is the number of the neuron input; f is the neuron activation function; w is the input weight of a neuron; θ is the neuron activation level.

The problem solving efficiency depends on the choice of the neural network architecture and its learning. The choice of the optimal architecture comes down to finding a network that solves the problem with a minimum target error:

$$E(w) = \frac{1}{2}\sum_{j=1}^{p}(y_j - d_j),\qquad(17)$$

where y_i is the value of the j-th output of a neural network; d_j is the target value of the j-th output; p is the number of neurons in the output layer.

The following types of neural networks are most suitable for this: a multilayer perceptron, a network with radial basic elements, a probabilistic network.

The neural network learns by the backpropagation method and the learning is similar to the learning by minimizing the target error (17).

The neural network learning algorithm is shown in Fig. 9.

The algorithm includes the following steps:

1. Feed the input vector to the neural network input (8) and calculate the values of the neural network outputs using system (13).
2. Calculate the auxiliary variable for the neural network output layer:

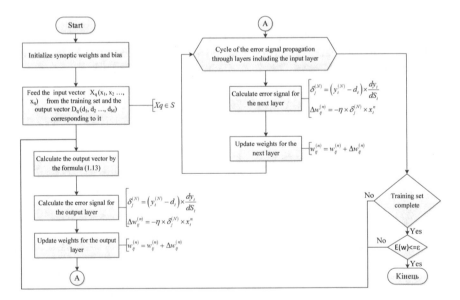

Fig. 9 Algorithm of the neural network learning

$$\delta_j^{(N)} = \left(y_j^{(N)} - d_i\right) \times \frac{dy_i}{dS_i}, \qquad (18)$$

where $y_j^{(N)}$ is the value of the ith output of the neural output layer; S_i is the weighted sum of outputs; N is the number of layers.

Change the weights $\Delta w_{ij}^{(N)}$ of the i-th layer of the j-the neuron:

$$\Delta w_{ij}^{(N)} = -\eta \times \delta_j^{(N)} \times x_{ij}^n, \qquad (19)$$

where η is the parameter that defines the rate of learning; x_{ij}^n is the value of the i-th input of the n-th layer neuron.

3. Calculate:

$$\delta_j^{(N)} = \left(\sum_k \delta_k^{(n+1)} \times w_{jk}^{(n+1)}\right) \times \frac{dy_i}{dS_i}, \qquad (20)$$

and (18) for other layers of the neural network as well, $n = N - 1 \dots 1$.

4. Update all the synapse weights of the neural network:

$$w_{ij}^{(n)} = w_{ij}^{(n)} + \Delta w_{ij}^{(n)}. \qquad (21)$$

5. Calculate the value of the correspondence index by the expression (17). If the value does not fall within the specified interval, go to step 1. Determining the indices of the significance criterion of the approximation error depends on the specific parameter of the problems being solved, and the approach to determining the criteria indicators will be further proposed.

3 Multi-agent Subsystem of Data Collection and Storage

Dramatic changes are taking place in the monitoring systems of the network infrastructure, caused by the competition in the market, increased requirements for the quality of security, technical re-equipment of communication networks, and changes in traffic distribution. All this leads to the need to control a large number of parameters of various networks operation.

To achieve this, a data collection and storage subsystem is used in the network monitoring system. It changes the concept of the operation system, as it does not collect data on parameters of individual stations but operational parameters of the entire network, and also automates many routine processes for collecting and processing network data.

The analysis of this information enables identifying various threats and failures, such as:

- network hacking that have not been detected by classical perimeter protection means (IPS/IDS);
- spread of viruses, "worms" and spyware that have not been detected by standard anti-virus tools;
- incorrect user actions (for example, large-scale downloads from torrent trackers, access to network segments to which there is no access, an attempt to access confidential information, etc.);
- new devices on the network and their behaviour;
- errors in equipment operation;
- bottlenecks in the network and other possible failures.

The architecture of subsystems designed to solve problems of collecting and storing parameters received from sensors are characterized not only by their target functions but also by operational capabilities that enable implementing target functions, by the hierarchy and degree of parallelism in solving problems, the homogeneity or heterogeneity of the modular structure, by collecting information in the real time mode, data processing and network exchange of information with subscribers [26].

In addition,

- the operation of network equipment must not be intervened;
- statistical information must be constantly collected, which enables creating large-scale databases necessary for conducting pseudo-operational and statistical analysis of the network;
- high speed of processing requests to provide necessary information resources and services must be ensured;
- complete information about the state of all components of the telecommunication and information infrastructure of the network must be collected, processed, and stored in the real time mode, regardless of the network architecture, type of switches and supplier;
- a single standardized information center to store data on the state of the system and network must be created.

Because of the large volume of events that go along the diagnostic monitoring, the variety of events and devices in an open system to diagnose and the need for real-time operation, taking into account the high variability of the external environment, the task of building a diagnostic sensor network must be considered as a problem of big data processing [26]. Solving this problem is associated with the implementation of new programming paradigms that support the capability of distributed interaction of autonomous active devices while solving a specific operational task [27].

To solve the above problems, the most appropriate technology is to implement a multi-agent system using autonomous software agents.

3.1 The Structure of a Multi-agent Data Collection and Storage Subsystem

To identify various threats and failures, the monitoring system must monitor a large number of parameters of network components states, which is implemented at different levels [28].

- Data channel level that defines the methods of access to the data transmission medium and enables transmitting a data frame between any nodes in typical-type networks according to the physical address of the network device. The addresses used at the data channel layer in local networks are often called MAC-addresses (MAC—media access control).
- Network layer that provides data delivery between any two nodes in a network with an arbitrary topology, while reliable data delivery from the sending node to the receiving node is not ensured. At this level, such functions as routing logical addresses of network nodes, creating and maintaining routing tables, fragmentation and data collection are performed.
- Session level that provides managing session, dialogues as well as synchronization within the messaging procedure, error control, transaction processing, support for calling RPC remote procedures.
- Applied level that is a collection of network services for ultimate users and applications. The examples of such services are e-mail messaging, file transfer between network nodes, network node management programs.

To study and analyse information, an open access to local and remote information sources is needed. But a data integration problem also arises. Various collections of state parameters of network components, even located on the same physical node, often have different logical inputs and do not provide through connection of data from different sources. The need to take into account all available information on a particular issue requires that the data collection and storage subsystem provide transparent means of access to distributed information for users.

One of the options to solve the problem is to restructure heterogeneous data. To restructure the monitoring data, it is necessary to harmonize information that can be compared to a single view or data structure expansion by adding unique features that make it possible to compare data from different sources. In some way or another, this task is very time consuming, especially for remote storage facilities since it requires technical and organizational interaction from sources. In addition, after the data restructuring, a qualitative modification of the means of access to information is required considering new data structures.

Therefore, a model of a data collection and storage system is proposed that provides work with a variety of heterogeneous sources by integrating them in order to obtain a more complete collection of related information. The system is based on a multi-agent approach and enables processing requests continuously while modifying the set and structures that are used in databases.

Depending on the problem being solved, the implementation of a multi-agent system can change dramatically, however, software agents retain the properties of intermediary activity: agents constantly interact with users or other programs.

For interaction between all agents, it is proposed to use a group of intelligent request agents whose purpose is to coordinate information collecting agents, restructure the obtained information and implement protocols and message transfer mechanisms between all the agents of a model.

The functions of each data collecting agent include:

- collecting and accumulating data in a small intermediate storage;
- preliminary data processing in real time;
- correcting sampling periods;
- addressing request agents to conduct additional measurements and a comprehensive analysis of the situation.

The data collection and storage model is given in Fig. 10.
The data collection and storage model comprises the following elements:

- switch agent and network agent that collect data from the first two layers described above. Since the functioning of a channel and network layer is provided mainly by active network equipment and is usually implemented by the components such as network adapters, repeaters, bridges, hubs, switches, routers, these agents operate on the basis of SNMP protocol to minimize the intervention in the operation of network equipment. MIB files are used as a staging warehouse. The task of these agents is to standardize data from files for further transmission to request agents. Also, agents are assigned the task of managing the emergency message delivery, since SNMP protocol works through unreliable UDP protocol [29];
- session agent that provides collecting data on a username, the name of terminal line, the astronomical time of the session start, the length of inactivity of the

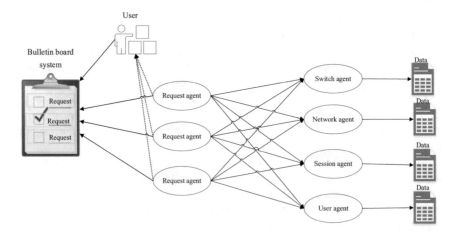

Fig. 10 The data collection and storage model

terminal line since the last exchange, the identifier of *shell* command interpreter for all users who work in the system. Staging warehouses might differ depending on the operating system. For example, for UNIX systems, such storage is system files /etc./utmp, /etc./wtmp, /etc./inittab;

- application agent is responsible for collecting data from various applications specific to a particular information and computing system;
- request agents aimed at processing requests for selecting data from users of the collection system, coordinating other agents to collect the necessary data as well as restructuring the obtained data to store statistical data about the system as a whole.

3.2 The Model of a Multi-agent Data Collection and Storage Subsystem

A multi-agent system (MAS) is a system which consists of a set of agents that have specific roles and interact with each other to solve problems that fall out of the capabilities or knowledge of an individual agent [30]. It is based on the transition from passive elements, which are described as classes of objects, to active ones, which are described as agents or models that operate actively, have autonomous behaviour, can make decisions following a certain set of rules, can interact with the environment and with other agents, can act to achieve set goals, and can change as well [30].

To design MAS, standard FIPA is used, this standard regulates MAS general architecture, the methods of interaction between agents, and the agent life cycle. In the context of this standard, request agents Ag_{select} perform the functions of AMS and DF, which allows agents to interact between one another. This service must necessarily provide agents with at least "white pages", where agents can find out the addresses of other agents.

Agent Management System (AMS) is an agent that regulates the entire platform. This agent is necessary to allow the platform agents to interact, it enables their access to the "white pages" service and regulates their life cycle. AMS provides various platform operations, such as: creating and deleting agents, deleting containers, shutting down the platform. Each agent needs to register with AMS to receive a personal AID. The AMS agent runs inside the main platform container but can exist in any container. An agent performing actions related to the platform operation must initially ask the AMS agent for confirmation.

Directory Facilitator (DF) is an agent that provides service registration and search for a service agent in yellow pages. Platform agents can subscribe to DF agent to receive information about the registration of the required service.

To describe a multi-agent data collection and storage system, a set of abstract agents is defined. So, a multi-agent system can be defines as follows:

$$MAC = (Ag_{abstr}, Ag_{select}), \tag{22}$$

where $Ag_{abstr} \in \{Ag\}$ is a set of abstract agents $\{Ag^1,...,Ag^4\}$, instances of switch agents, network agents, session agents and application agents that can be dynamically involved into the system;

$Ag_{select} \in \{Ag\}$ is a specialized agent that enables service registration and agent search, as well as regulates the interaction of platform agents and provides agents with access to the service of the white pages of the platform and controls their life cycle.

The objects of finite set (22) create *MAC*. Solving any task in the context of *MAC* starts with the initialization of one of the system agents:

$$MAC_{init} = (Ag_{init}, Ag_{sectinit}),$$
$$Ag_{init} = \{Ag_1^1, \ldots, Ag_{h_1}^1, \ldots, Ag_1^4, \ldots, Ag_{h_n}^4\}, h_1, \ldots, h_n \in \mathbb{N}\, i$$
$$\forall Ag_j^i \in Ag_{init} : Ag^i :\Leftrightarrow Ag_j^i \wedge Ag^i \in Ag_{init}. \tag{23}$$

Agent Ag_j^i inherits the behaviour and all the initial knowledge of the prototype agent, and can also have some additional features or knowledge (for example, a unique identifier that allows other agents of the multi-agent system to distinguish from others and use it as an address for communication).

MAC environment is open and within it there are agents that had been already initialized at the moment the first agent ran. Therefore agents from Ag_{init} set can be augmented by the agents that have already existed in the above open environment.

Hence, a set of similar available agents can be added to the model:

$$Ag_{existent} = \{Ag_1^1, \ldots, Ag_{h_1}^1, \ldots, Ag_1^m, \ldots, Ag_{h_n}^m\},$$
$$Ag^m \notin \{Ag\} \wedge Ag^m \notin Ag_{abstr}. \tag{24}$$

A software designer who develops Ag_{abstr} needs information on $Ag_{existent}$, since Ag_{abstr} use services of these agents.

MAC operation bases on the agents' interaction. New agents can be added to the system at any time, while old agents can be deleted from it. Every agent has a unique address that can be registered by Ag_{select} and become accessible for all *MAC* agents. All agents know the address of Ag_{select} agents from the start.

The software implementation of a multi-agent data collection and storage system is a system, as a single environment, that consists of autonomous modules with decentralized control. The main goal of such a model is to get the advantages of the decentralized management, namely flexibility, resistance to negative environmental impacts, and efficient use of available resources.

3.3   The Interaction Protocol of ShS Monitoring Agents

Within a unified and integrated execution model, an agent can provide one or more services, which may include access to the external software, users, and communication capabilities of the system.

The messaging service is one of the main elements of the multi-agent system architecture. The service is based on asynchronous messaging. Each agent has its own "mailbox"—a queue of incoming messages that contains all messages sent to the agent. The moment a message enters the incoming message queue, the agent gets a notification.

Consider the objectives of the components of a multi-agent data collection system. The request agent is the interface between the operator \ administrator, the data warehouse, and the network nodes where the instances of collection agents run. The instances of collection agents of various levels are the interface between the request agent and sensors for monitoring parameters. It is a well-defined architecture a client–server interaction model. In this system, the roles of a client and server are somewhat ambiguous.

For example, an instance of the session agent is a kind of service that runs on a node (that is monitored) and processes requests on a specific TCP port or protocol and acts as a server while the request agent acts as a client that addresses the collection agent server. Requests (more precisely messages) can be also sent from collection agents to request agents. When such a message is sent, the collection agent and the request agent swap roles. It means that a request agent in this case must be a server that operates on UDP/TCP port, but a collecting agent is a client.

All system agents operate at layer 7 of the OSI model, that is, are an application layer service. The interaction of request agents and collection agents is organized using Protocol Data Messages (PDM), which are encapsulated in the transport protocol. At the same time, each PDM message contains a specific command (to read data from the staging warehouse, change the collection agent configuration, or the response generated by the collection agent). In general, the interaction of system agents can be represented in the following sequence: request agent → PDM-UDP/TCP-IP-Ethernet-IP-UDP/TCP-PDM → collection agent. To ensure reliability, PDM data can be encrypted.

For the exchange between the agent and the manager, a specific set of PDM command messages is used:

- Trap is a one-way message from the collection agent to the request agent on any event, for example, deviations of any monitoring parameter from a certain range.
- GetReponse is a response from the collection agent to the request agent that returns information on data that have been requested for.
- GetRequest is a request to the collection agent from the request agents that is used to obtain a value from the staging warehouse. використовується для отримання значення з проміжного сховища даних.
- SetRequest is a request to the agent to configure the parameters of the agent itself.

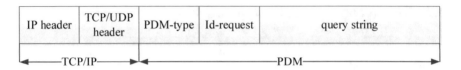

IP header	TCP/UDP header	PDM-type	Id-request	query string

←————TCP/IP————→←————————————————PDM————————————————→

Fig. 11 PDM package structure

- GetBulkRequest is a request to the agent to synchronize the data from the staging warehouse with the main database.
- InformRequest is a one-way messaging between request agents. It can be used, for example, to exchange information on collection agents that are controlled by a request agent. In response, the request agent creates the package to confirm the fact that output data have been obtained.

All PDM (but Trap) are composed of a particular set of fields used to record necessary information shown in Fig. 11.

PDM package fields have the following value:

- PDM type has a digital request identifier (Get, Set, Responde, Trap);
- a request identifier is a set of symbols that integrate a request and a response into a whole;
- a request line is a sql-like line using which the relevant data are sampled or set.

There are some general features of the PDM protocol that must be considered. The receiver first tries to decode messages. If the receiver fails to decode the PDM, the packet is dropped without any actions. Next, the message is processed. If necessary, a response is generated as JSON with the specified request identifier.

Based on the results of the analysis of the structure of the multi-agent subsystem for data collecting and processing and the functions of agents, the agent architecture has been developed (Fig. 12).

The management module configures the parameters of the agent and data transfer from the staging database to the central database. It also authenticates the subject of interaction.

The data processing and standardization module receives parameters from a specific data source that characterizes the state of the ShS component, transforms them into the form required to store in the staging database. The received data are transferred at the input to the analysis module and written to the staging database.

The analysis module, depending on the agent type, assesses the state of the corresponding resource operation and, if necessary, transfers information to the event module.

The event module, depending on the type of agent and settings, can provide information about the event, send an ICMP packet that informs the attacking node about the unavailability of a node, network or service.

The message module, depending on a request, generates a message that contains settings, a message that requests for certain information or a message that initiates the process of data transfer from the staging database to the central database.

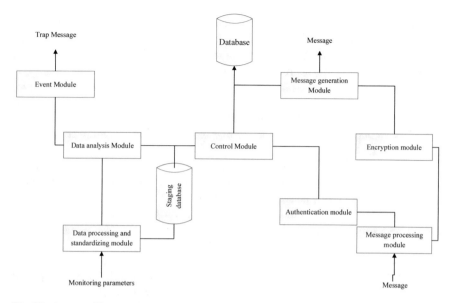

Fig. 12 Agent architecture

The message module, depending on the received message, transfers the received settings to the management module, sends a request to the control module to send events or ordered data in response to the request.

The encryption module is used to encrypt messages that are exchanged between agents.

The authentication module is used to identify and authenticate other agents.

4 Formulating Approaches to Solve the Problems of classification of Anomalies in Self-healing Systems

When solving complex problems of classification, regression, forecasting, it often happens that none of the algorithms provides the desired quality of dependence recovery. In such cases, it is reasonable to build compositions of algorithms where the errors of individual algorithms are compensated. Along with the set of objects and the set of values of the objective function corresponding to them, an auxiliary set called the estimation space is introduced. Algorithms are considered where a function called the algorithmic operator establishes a correspondence between a set of objects and the estimation space, and a function called the decision rule establishes a correspondence between the estimation space and a set of the objective function values. Thus, the considered algorithms are of the form of a superposition of an algorithmic operator and a decision rule. Many classification algorithms are of the following structure: first, the assessments of the object belonging to classes are calculated, and then the decision rule transforms these assessments into the class number. Figure 13 illustrates the general algorithmic composition scheme. The assessment value can be the probability of the object belonging to the class, the distance from the object to the shared surface, the degree of classification confidence, and so on.

There are several well-known methods of combining basic algorithms in a composition: voting, weighted voting, a mixture of experts. These methods are often used when basic algorithms differ significantly from each other. In cases when it is necessary to build a composition using one basic algorithm, bagging or bootstrap aggregation are widely used [31].

The idea behind bagging is that the basic algorithm permanently learns by random subsamples with repetitions by the training sample. This method of generating subsamples is commonly called bootstrap. The random subspace method (RSM) is similar to bagging [32]. Its idea is to create variability in learning by selecting random subsets. A well-known example of the use of bagging and RSM is RandomForest [33].

Another well-known way to combine basic algorithms into an ensemble is boosting. The idea of boosting lies in greedy selection of the next algorithm to add to the composition so that it compensates for the errors at this stage in the best way. The widely known examples of boosting are AdaBoost [34] and Gradient boosting [35].

David Wolpert firstly proposed stacked generalization in a general way in 1992 in his work [35]. The basic idea of stacked generalization is to use basic classifiers to get predictions and use them as features for a "generalizing" algorithm. In other words, the main idea of stacking is to transform the initial feature space into a new space whose points are the prediction of basic algorithms. It is proposed to first select a set of pairs of arbitrary subsets from the training set, then, train basic algorithms for each pair at the first subset and due to them predict a target variable for the second subset. The predicted values become objects of the new space. In particular, the author considers the case of choosing all possible pairs of subsets, where the second subset comprises a single object while the first subset consists of the entire learning set except for the object from the second subset (leave-one-out procedure). Obviously, this method enables assigning each point of the initial feature space to a point in the new space. The author generalizes the stacking idea proposing to train

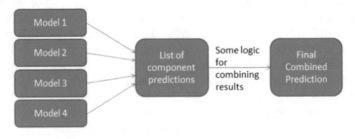

Fig. 13 General algorithmic composition scheme

the basic classifiers (of the first level) by meta-features (of the first level) to get meta-features of the second level, and so on.

4.1 Principles of Sequential Training of Basic Algorithms

Voting procedures can be of various forms such as absolute majority voting, majoritarian electoral systems that approve voting, and many others.

The process of sequential training of basic algorithms is used most often when constructing compositions. Let us first consider this process in its most general form, which is illustrated by the algorithm presented in Fig. 14.

Parameters X^l, Y^l, are a training set, μ is the method of basic classifiers training, T is the maximum number of algorithms in a composition, $F(b_1,..., b_t)$ is a algorithmic composition, $M(W^l)$ is the function of classifier weights modification.

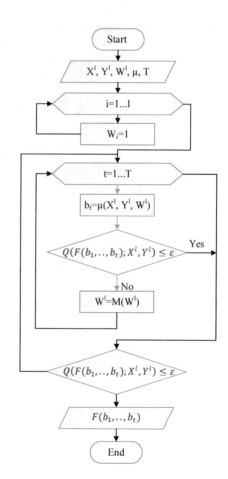

Fig. 14 Building an algorithmic composition by sequential training of basic algorithms

At the first stage, using the standard learning method μ, the first basic algorithm b_1 is built. If its quality meets the requirements, then there is no necessity to build the composition further and the process ends there. Otherwise, algorithm b_1 is fixed and the second algorithm b_2 is built, while the correcting operation F is being simultaneously optimized. At the t-th step, the basic algorithm b_t and the correcting operation F are optimized for fixed $b_1, \ldots, b_{t-1}$:

$$b_1 = \arg \max_{b} Q(b; X^l, Y^l); \tag{25}$$

$$b_2 = \arg \max_{b,F} Q(F(b_1, b); X^l, Y^l); \tag{26}$$

$$b_t = \arg \max_{b,F} Q(F(b_1, \ldots, b_t); X^l, Y^l); \tag{27}$$

In many cases, to solve problem (27), standard training methods can be used that can solve a simpler problem (25). To do this, modified vectors of weights W^l and responses Y^l are fed to the input of a standard training method $\mu(X^l, Y^l, W^l)$. The specific modification method depends on the type of classification or regression problem and the type of correcting operation F. For each individual case, the formulas for recalculating the weights and responses are derived separately. Weights modification is usually reduced to increasing the weight of the "heaviest" objects, where the previous basic algorithms have made more errors, and the modification of responses is reduced to approximating the computational error $y_i - a(x_i)$ instead of approximating output responses y_i.

Basic algorithm b_t, optimal at the t-th step is not optimal after adding the next algorithms. Process (25)–(27) can be generalized by alternating adding new algorithms with re-setting previous algorithms:

$$b_k = \arg \max_{b,F} Q(F(b_1, \ldots, b_{k-1}, b, b_{k+1} b_t); X^l, Y^l), \quad 1 \le k < t. \tag{28}$$

As regards solution methods, this task is not much different from the task of building the last basic algorithm (27).

The stopping criteria can be used differently, depending on the task specificity, several criteria can be used if:

- a given number of basic algorithms T have been built;
- the specified accuracy has been achieved in the training set:

$$Q(F(b_1, \ldots, b_t); X^l, Y^l) \le \varepsilon; \tag{29}$$

- the achieved accuracy in the control sample X^k cannot be improved within the last d steps: $t - t^* > d$, where d is an algorithm parameter;

$$t^* = \arg \max_{s=1,\ldots,t} Q(F(b_1, \ldots, b_s); X^k, Y^k). \tag{30}$$

Meeting this criterion is considered a sign of overfitting. As the final decision, the composition built on the t^*-th step is taken.

Basic algorithms can be built easier when the correcting operation does not have its own adjustable parameters. These methods include voting by majority and seniority.

4.2 Principles of Algorithms Combination by Majority Vote

Consider a classification problem with two classes $Y = \{-1, +1\}$ by the assessment space $R = \mathbb{R}$ and the decision rule $C(b) = sign(b)$. Regard simple voting as a corrective operation. Consider the composition quality functional a, that is equal to the number of learning errors:

$$Q(a; X^l) = \sum_{i=1}^{l} [y_i a(x_i) < 0] = \sum_{i=1}^{l} [y_i b_1(x_i) + \cdots + y_i b_j(x_i)]. \qquad (31)$$

Figure 15 shows an algorithm of composition building by majority voting.

Parameters X^l, Y^l are a training set, μ is the method of basic classifiers training, T is the maximum number of algorithms in a composition, $F(b_1,..., b_t)$ is an algorithmic composition, l_l is a training set length, $Sort(X^l, M_i)$ is the function of ordering X^l by indent values M_i.

If an indent is negative $M_{it} < 0$, the composition of the first t basic algorithms makes errors on the object x_i. To compensate for composition errors, the basic algorithm b_t + 1 is not trained on the entire sample X^l but on the objects with minimum values M_{it}. Or, the functional $Q(b_{t+1}; W^l)$ is minimized by the object weights $w_i = [M_{it} \leq M_0]$.

M_0 must be selected so that the training set contains not too small objects (otherwise basic algorithms of too low quality will be built), but also not very large ones (otherwise, almost identical algorithms will be built). That is why instead of M_0, it is better to introduce the parameter of learning subsamples length l_1 and find its optimal values.

4.3 Principles of a Combination of Algorithms by Seniority Voting

Consider the classification problem with an arbitrary number of classes $|Y| = M$, where $R = \{0, 1\}$. Basic algorithms are used that classify into two disjoint classes $b_t: X \rightarrow \{0, 1\}$.

Let us consider what is happening when the basic algorithm b_t is added. If $b_t(x) = 1$, it is said that "b_t singles out object x" and object x belongs to class c_t. Let U_t defines a subset of samples that have not been singled out by any of the previous basic algorithms:

Fig. 15 Algorithm for
building composition to vote
by majority

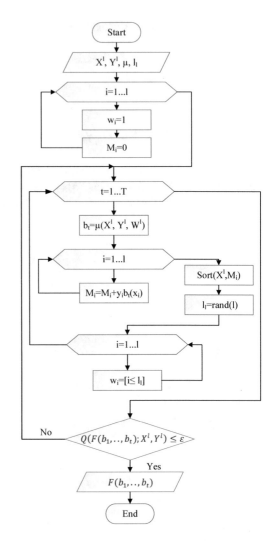

$$U_t = \left\{ x_i \in X^l : b_1(x_i) = \cdots = b_{t-1}(x_i) = 0 \right\}. \tag{32}$$

Figure 16 shows the algorithm to build a seniority voting composition.

Parameters X^l, Y^l are a training set, μ is the method to train basic classifiers, T is the maximum number of algorithms in a composition, $F(b_1,\ldots,b_t)$ is an algorithmic composition, λ is a penalty for refusing to classify, Z^l is a binary vector of responses.

If $x_i \in X^l \backslash U_t$, object x_i is already classified, and value $b_t(x_i)$ does not affect the composition response $a(x_i)$. For such objects, the weight w_i is equal to zero.

If $x_i \in U_t$ i $y_i \neq c_t$, the basic algorithm error $b_t(x_i) = 1$ results in the error of the entire composition: $a(x_i) = c_t \neq y_i$. The weight of such objects equals 1: $w_i = 1$.

Fig. 16 Algorithm for
building a composition to
vote by seniority

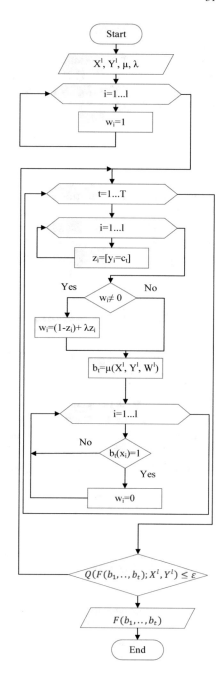

If $x_i \in U_t$ i $y_i = c_t$, the basic algorithm error $b_t(x_i) = 0$, results to the situation when the composition of the first t algorithms refuses from the classification x_i. This error can still be corrected by the next algorithms. Therefore, for such objects $w_i = \lambda$, where $\lambda \in [0, 1]$ is a penalty of the refusal. So, the minimization of the functional $Q(a) = Q(F(b_1, ..., b_t))$ according to the basic algorithm b_t is equivalent to minimizing the functional

$$Q(b_t) = \sum_{i=1}^{l} w_i[b_t(x_i) \neq [y_i = c_t]]. \tag{33}$$

This is a standard functional of the weighted number of errors. To minimize it, let us use the standard method of training $b_t = \mu(X^l, Z^l, W^l)$, giving it the vector of objects weight $W^l = (w_1, ..., w_l)$ and the binary vector of responses $Z^l = (z_1, ..., z_l)$, where $z_i = [y_i = c_t]$. As a result, the basic algorithms b_t tends to select as many *own* class objects c_t and as less *foreign* objects.

4.4   Principles of Combining Algorithms by Boosting Methods

Boosting is a method aimed at transforming weak models into strong ones by building an ensemble of classifiers.

Within boosting, the classifiers are sequentially trained. Thus, the training dataset at each next step depends on the prediction accuracy of the previous base classifier.

Figure 17 shows the scheme of boosting.

Consider the task of classification into two classes $Y = \{-1, +1\}$. Assume that basic algorithms also give just two responses -1 and $+1$, and the decision rule is fixed: $C(b) = sign(b)$. The target algorithmic composition is as follows:

$$a(x) = C(F(b_1(x), ..., b_r(x))) = sign\left(\sum_{t=1}^{T} \alpha_t b_t(x)\right), x \in X. \tag{34}$$

Let us define the functional of composition quality as the number of errors made by it on the training set:

$$Q_T = \sum_{t=1}^{t} [y_i \sum_{t=1}^{T} \alpha_t b_t(x_i) < 0]. \tag{35}$$

To simplify the task of Q_T functional minimization, two heuristics are introduced.

Heuristic 1. If addend $\alpha_t b_t(x)$ is augmented into the composition, only basic algorithm b_t and coefficient α_t, are being optimized but not the previous addends $\alpha_j b_j(x)$, $j = 1, ..., t - 1$ [37].

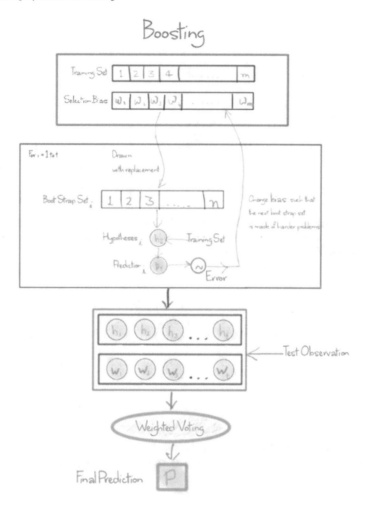

Fig. 17 Scheme of boosting to build composition

Heuristic 2. To solve the task of optimizing parameter α_t analytically, the boundary loss function in the functional Q_t is approximated by some of its continuously differentiable upper values [37].

The second heuristic is widely used in the classification theory. In particular, the logarithmic function is associated with the maximum likelihood principle and is used in neural networks and logistic regression. The piecewise linear function approximation is related to the principle of optimal shared surface (maximization of the spacing between classes) and is used in the support vector machine. The examples of other approximations are shown in Fig. 18.

The process of sequential training of basic algorithms within exponential approximation results in the well-known algorithm [8].

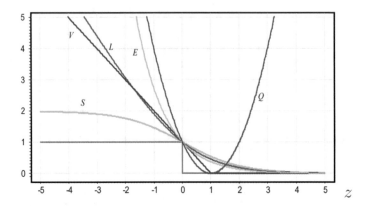

Fig. 18 Smooth upper approximations of the threshold loss function [z < 0]: S(z) = 2(1 + ez) − 1—sigmoid; L(z) = log2(1 + e − z)—logarithmic; V(z) = (1 − z)+—piecewise linear; E(z) = e − z—exponential; Q(z) = (1 − z)2—squared

Let us evaluate Q_T:

$$
Q_T \le \tilde{Q}_T = \sum_{i=1}^{l} \exp\left(-y_i \sum_{t=1}^{T} \alpha_t b_t(x_i)\right)
$$

$$
= \sum_{i=1}^{l} \underbrace{\exp\left(-y_i \sum_{t=1}^{T-1} \alpha_t b_t(x_i)\right)}_{w_i} \exp(-y_i \alpha_T b_T(x_i)). \tag{36}
$$

The introduced objects weights w_i do not depend on α_T and b_T, so, w_i can be calculated before building the basic algorithm b_T. Let us introduce the normalized vector

$$
\tilde{W} = \left(\tilde{W}_1, \ldots, \tilde{W}_i, \ldots, \tilde{W}_n\right), \tag{37}
$$

where $\tilde{W}_i = W_i \big/ \sum_{j=1}^{i} W_j$.

Let Q be the standard quality functional of the classification algorithm b on the training set X^l, Y^l with the normalized vector of weights of the objects U^l:

$$
Q(b; U^l) = \sum_{i=1}^{l} u_i[y_i b(x_i) < 0], \quad \sum_{i=1}^{l} u_i = 1. \tag{38}
$$

Let $\min_{b} Q(b; W^l) < 1/2$ for any normalized weight vector W^l. Then $\tilde{Q}_T$ functional minimum is made if:

$$b_T = \arg \min_b Q\left(b; \widetilde{W}^l\right). \tag{39}$$

$$b_T = \frac{1}{2} \ln \left(1 - Q\left(b_T; \widetilde{W}^l\right)\right) \Big/ Q\left(b_T; \widetilde{W}^l\right). \tag{40}$$

Figure 19 shows AdaBoost—building a linear combination of classifiers.

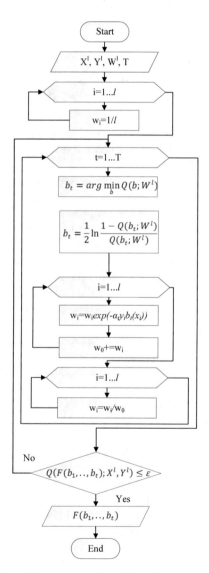

Fig. 19 AdaBoost—building a linear combination of classifiers

Parameters X^l, Y^l are a training set, μ is the method of basic classifiers training, T is the maximum number of algorithms in a composition, $F(b_1, ..., b_t)$ is an algorithmic composition.

After building a number of basic algorithms (for example, a couple of dozen), it is reasonable to analyse the distribution of object weights. The objects with the highest weight w_i are most likely noise burst that must be excluded from the sample, and then start building the composition once more. In general, boosting can be used as a universal method to filter bursts before using any other method of classification.

4.5 Principles of Combination of Algorithms by the Bagging Method and the Method of Random Subspaces

The basic algorithms that make up a linear combination must be sufficiently different so that their errors compensate for each other. It is not reasonable to build the same or nearly identical algorithms.

In boosting and other sequential algorithms, the difference is achieved by recalculating the object weights. But another strategy to increase the difference is also possible, when the basic algorithms are adjusted independently of each other on randomly selected subsets of the training set or on different random subsets of features. Another way to ensure differences is to choose random initial approximations while optimizing a vector of parameters (this is usually done while adjusting neural networks), or while using stochastic optimization algorithms. The obtained basic algorithms are combined into a composition using simple or weighted voting. In the case of weighted voting, standard linear methods are used to adjust the coefficients α_t.

The method of bagging (or bootstrap aggregation) was proposed by Leo Breiman in 1996 [37]. Various training subsets of the l length are built form the initial training set and the same length l by bootstrap, that is random selection with returns. At the same time, some objects get into a subset several times, while others—never. It can be said that a part of objects that belong to each subset tends to $1 - e^{-1} \approx 0.632$ at $l \to \infty$. Basic algorithms having been trained on subsets are combined into a composition by simple voting.

Figure 20 shows the general scheme of bagging.

The efficiency of bagging is due to two properties. First, due to different basic algorithms, their errors are mutually compensated for when voting. Secondly, outliers can fail to get in some training subsets. Then the algorithm built on the basis of the subset can be more accurate than the algorithm built on the basis of the entire set. Begging is especially effective on small sets, when the exclusion of even a small part of training objects results in building significantly different basic algorithms.

In the case of very large redundant sets, subsets of shorter length $l_0 < l$ must be built, and in this case there is a problem of choosing the optimal value of l_0.

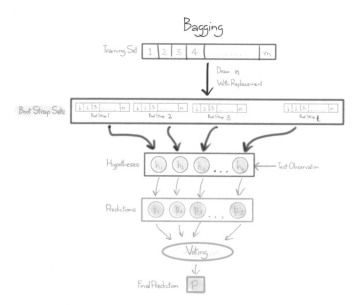

Fig. 20 Scheme of bagging to build a composition

In the random subspace method (RSM), the basic algorithms are trained on different subsets of the feature space, and are also selected at random [38]. This method is better to solve problems with a large number of features and a relatively small number of objects as well as when there are redundant uninformative features. In these cases, algorithms built by a part of the feature space can show more accurate results than algorithms built by all features.

The generalization of bagging and RSM results in the algorithm that is shown in Fig. 21.

Parameters X^l, Y^l is a training set, μ is the method to train basic classifiers, T is the maximum number of algorithms in a composition, $F(b_1,..., b_t)$ is an algorithmic composition, l' is a training set length, n' is the length of the features space, ε_1 is the threshold for the basic classifiers quality during training, ε_2 is the threshold for the basic classifiers quality during control.

Let objects be described by a set of features $F = \{f_1, ..., f_n\}$ and the training method $\mu(G, U)$ is building an algorithmic operator by a training subset $U \subseteq X^l$, using just a part of features space $G \subseteq F$. Numbers $l' = |U| \le l$ and $n' = |G| \le n$ are the method parameters. The algorithm in Fig. 3.7 corresponds to bagging when $G = F$, and to the method of random subsets when $U = X^l$.

Steps 5–6 filter poor basic algorithm. Basic algorithm b_t is considered poor if its quality at the training subset is worse than the given threshold ε_1, or if the quality on the test data $X^l \backslash U$ is worse than ε_2. It might happen when there are too many noise outliers in the training set or when there are too few informative features.

None of the methods to build liner compositions is definitely the best. Each method is superior to others in solving certain applied problems.

Fig. 21 Generalized
algorithm of bagging and
RSM

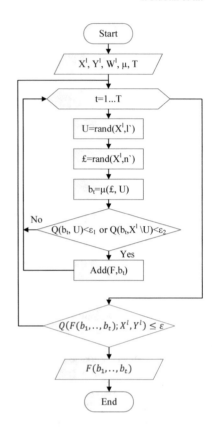

4.6 Principles to Combine Algorithms Bystacking

Stacking is another method to combine classifiers that introduces the notion of training meta-algorithm. Unlike bagging and boosting, while stacking the classifiers of different kinds are used. The idea of stacking is to [36]:

(1)   split the training set into two subsets that do not overlap;
(2)   train several basic classifiers on the first subset;
(3)   test basic classifiers on the second subset;
(4)   train the meta-algorithm using the predictions from the previous paragraph as input, and the real classes of objects as an output.

Figure 22 shows the scheme of stacking.

Consider the task of regression. Let all K of basic models be $f_k(x)$ of regression algorithms. The resulting model is built as follows:

$$f(x) = \sum_{k=1}^{K} w_k f_k(x). \tag{41}$$

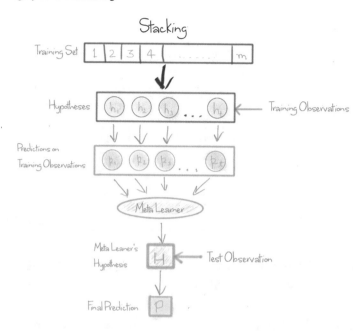

Fig. 22 Scheme of stacking to build a composition

The weight values are determined as follow:

$$\hat{w} = \arg\min_{w} \sum_{i=1}^{N} L(y_i, \sum_{k=1}^{K} w_k f_k(x_i)). \qquad (42)$$

But such a way leads to overfitting. Therefore, weight must be found using cross-validation, that is, a set is divided into M parts. Let *fold* (i) be the part that contains the i-th object and $f_k^{-fold(i)}$ be the algorithm that have been trained on all the folds but *fold* (i), then:

$$\hat{w} = \arg\min_{w} \left(\sum_{i=1}^{N} L(y_i, \sum_{k=1}^{K} w_k f_k^{-fold(i)}(x_i) \right). \qquad (43)$$

To reduce overfitting, the weight positiveness conditions can be added to the regularizer functional:

$$\lambda \sum_{k=1}^{K} \left(w_k - \frac{1}{K} \right)^2. \qquad (44)$$

It must be noted that stacking does not always significantly improve the quality of the best of the basic algorithms.

Table 1 TCP parameters

Parameter	Description
Duration	Connection duration (seconds)
Protocol_type	Transport layer protocol
Service	Application layer service
Number of data bytes source—destination	Input stream, byte
Number of data bytes destination—source	Output stream, byte
Flagstatus	TCP-packet header flags
Land	Addresses coincide, 0 otherwise
Wrong_fragment	Number of wrong fragments
Urgentpackets	Urgent data in a packet (URG flag)

5 Method of Classification of Self-healing Systems State Based on Statistic Parameters

5.1 Setting the Problem of Classifying the State of the ShS Core Network

The classical task of object classification is suggested in article [8].

It is needed to develop an algorithm that can give the value of the target variable based on the features of a new object.

The dataset, which is used in the work to build a model, was modelled and obtained in the data exchange network of a cluster supercomputer. The set of parameters presented at the KDD Cup 2009 competition on machine learning competition was taken as a basis, but the parameters to monitor the data warehouse being added.

Tables 1, 2, 3 and 4 give the example of output data for the classification task.

5.2 Method to Classify the State of the ShS Core Network Based on a Modified Stacking Algorithm

Stacking uses the concept of meta-training, that is, it tries to train each classifier using an algorithm that enables identifying the best combination of basic models outputs [36].

The sequence of this algorithm operation in a simplified form consists of the following stages:

- Training set $X = \{x_1, x_n\}$ and a set of basic algorithms in the classification $A = \{a_1,...,a_m\}$ are fed to the algorithms input;
- X set is split into two subsets X^a and X^b that do not overlap;

Table 2 Session features

Parameter	Description
Hot	Number of "hot" indicators
num_failed_logins	Number of unsuccessful login attempts
logged_n	Successful login
num_compromised	Administrator access
root_shell	Number of administrator access attempts
num_root	Access check file operations
num_file_creations	Number of file creation operations
num_access_files	Number of access control file operations
num_compromised	Number of compromised statuses
is_hot_login	Client belonging to "hot" list
is_guest_login	Guest system sign

Table 3 Statistics of connections per 2 s

Parameter	Description
Count	Number of connections to matching host
serror_rate	Percentage of "SYN" error connection
Service	Percentage of"REJ" error connection/percentage of connections to the same initial port
rerror_rate	Percentage of connections to the same service
same_srv_rate	Percentage of connections to a different service
diff_srv_rate	Number of connections to different service
srv_count	Number of connections to the same service
srv_serror_rate	Percentage of "REJ" error connection
srv_rerror_rate	Percentage of error connection
srv_diff_host_rate	Percentage of connections to different hosts

Table 4 Features of cluster file system (Lustre)

Parameter	Description
num_exports	Number of exports to MDT including other Lustre services
Stats	Listed client connections on NID
lock_count	Number of locks
pool.granted	Luster distributed locking manager (ldlm) has granted locks
grant_rate	ldlm locked undo level called 'GR'
cancel_rate	Ldlm locked undo level called 'CR'

- A set of basic classifiers A is trained on subset X_a;
- Basic classifiers A are tested on subset X^b, as: $X^b \rightarrow Y_b$;
- Set Y^b is used as the initial data for a meta-algorithm while true values of a target variable are used as output values to train a meta-algorithm.

The algorithm of the classical stacking operation is presented in Fig. 23.

The drawback of this algorithm is that basic algorithms is not trained on the entire set of objects. This, in turn, leads to the fact that none of class k objects can get into the X^a training subset and a set of basic algorithms of A classifiers can fail to be trained to the full, and as a consequence, the entire stacking algorithm can fail to be trained as well.

The authors propose a modified stacking algorithm. The operation of this algorithm is described by the following tuple:

$$(A_{ij}, Out_{ij}, In_{ij}, x_n, A_{meta}), \tag{45}$$

where $A_{ij-1} : In_{ik} \rightarrow Out_{ij}$ can classify an arbitrary object of In_j set.

$$A_{meta} : In_{ik} \rightarrow Y, \tag{46}$$

can classify an arbitrary object of In_k set:

- x is a set of input values;
- k is a number of stacking levels;
- In_{ik} is a set of input objects of the i-th algorithm on the j-th level;

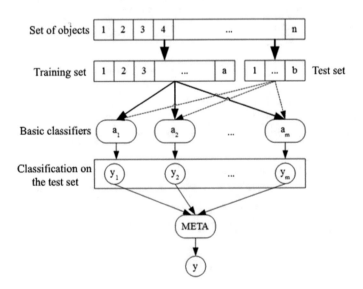

Fig. 23 Scheme of classical stacking

- *Out_{ij}* is a set of output objects of the *i*-th algorithm on the *j*-th level;
- *i* is an algorithm number on the level;
- *j* stacking level number;
- *Y* a set of target variables.

The operational algorithm comprises the following stages:

- Training set $X = \{x_1, ..., x_n\}$ and a set of basic algorithms of множина базових алгоритмів класифікації $A = \{a_1, ..., a_m\}$ classification are fed the algorithm input;
- Set *X* is split into *K* subsets that overlap. By uniform set *L* of return objects. Each subset is built using various objects of output set *X*. Approximately 37% objects remain outside the subset and are not used to build *K*-th subset.
- Set of basic classifiers *A* of the *K-th* level of stacking is trained on *K* subset;
- Basic classifiers of the *K-th* level are tested on the set of objects that do not get into the *K-th* subset;
- Set of objects that do not get into the *K-th* subset are used as initial data for a meta-algorithm while true values of the target variable as output data to train a meta-algorithm.

The modified algorithm of stacking operation is illustrated in Fig. 24.

This algorithm enables getting away from drawbacks before training and enables using it on small-size training sets.

5.3 Results of the Study of Methods to Classify the Network State

A modified stacking algorithm enables using fewer objects of a training set and leads to the stepwise reducing the feature space for a meta-classifier with less correlation.

The efficiency of the proposed algorithm has been compared with the operation of basic classifiers and classical stacking. To do comparative research, the dataset of the machine learning competition KDD 2009 was used as well as data obtained while monitoring the network infrastructure of the training data-center based of the network file system Lustre.

Tables 5 and 6 show the classes of attacks on the networks that belong to the training and test sets of KDD 2009.

To make the experiment, the following basic classifiers were used:

- kNN classifier;
- naive Bayes classifier;
- classification trees;
- SVM.

A multilayer perceptron was used as a meta-classifier.

The first stage of the experiment involves preparing data to train basic classifiers. Since such algorithms of machine learning as kNN and SVM (that are sensitive to

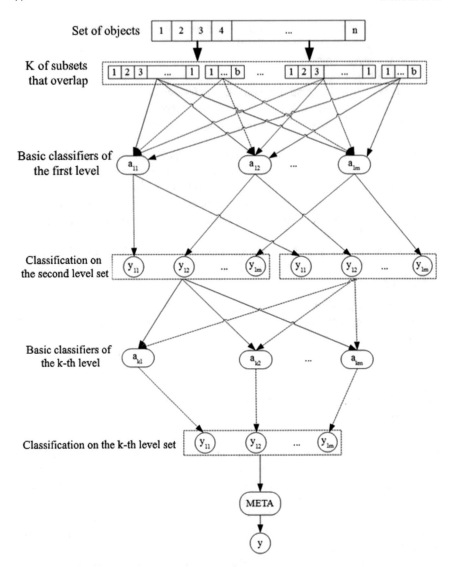

Fig. 24 Scheme of modified stacking

data scaling) are used in the experiment, minimax normalization is used for multiple features.

For categorical features, coding was used, which was considered in the article [37], and its essence is in the following.

Let F be a set of real functions, in which any natural number k equals one function of k variables. All the functions are symmetric, that is, for any function, the values of k variables with F are the same no matter the order of its arguments. An example

Table 5 Classes of attacks in the training and test sets

Training set	Test set
Back, buffer_overflow, ftp_write, guess_passwd, imap, ipsweep, land, loadmodule, multihop, neptune, nmap, phf, pod, portsweep, rootkit, satan, smurf, spy, teardrop, warezclient, warezmaster, normal	Guess_passwd, imap, ipsweep, land, loadmodule, multihop, neptune, nmap, phf, pod, portsweep, rootkit, satan, smurf, spy, teardrop, warezclient, warezmaster apache2, htttptunnel, mailbomb, mscan, named, perl, processtable, ps, saint, sendmail, snmpgetattack, snmpguess, sqlattack, udpstorm, worm, xlock, xsnoop, xtermbuffer_overflow, Neptune, warezmaster, smurf, normal

Table 6 Classes of attacks in the training and test sets

Training set	Test set
Back, neptune, pod, spy, normal	Back, neptune, pod, smurf, spy, teardrop, normal

of such large numbers can be the sum of arithmetic mean values, maximums, and so on.

To code the value f_{jj}-th of a categorical feature, a set of training objects with such a value is selected:

$$I = \{t \in \{1, 2, ..., l\} | f_{tj} - f_j\}, \tag{47}$$

a real feature is selected by which coding is performed, for example, s-th, and f_j value is coded by the value of the corresponding function with F, that is functions from $|I|$ variables from values f_{is}, For example, the protocol was coded by replacing it with the arithmetic mean value of session duration for such a category.

The second stage is to train basic algorithms, standard and modified stacking.

After that, an experimental study of the efficiency of the basic classifiers, classical and modified stacking was carried out, the results are illustrated in Tables 7 and 8.

6   The Technique to Monitor Self-healing Systems State

An increase in the amount of information processed by computing clusters, as well as a reduction in the number of service personnel, require the use of efficient means of monitoring computing resources, which leads to an increase in the number of parameters that such system must monitor. Due to the large data streams from various sensors, a system administrator might fail to track negative changes in the controlled parameters of the cluster computing network. To solve this task in the monitoring system, tools for automated expert data analysis based on machine learning are being widely introduced.

Table 7 Results of the experiment on KDD 2009 dataset

Number of tests	Number of correct solutions		Number of incorrect solutions			
	Training set	Test set	Errors of the 1st kind		Errors of the 2nd kind	
			Training set	Test set	Training set	Test set
Naive Bayes classifier	80.59	64.90	15.68	23.40	3.73	11.70
kNN classifier	85.40	70.40	10.60	19.97	4.00	9.63
SVM	84.30	66.47	10.60	19.93	5.10	13.60
Classification trees	85.70	72.80	8.10	15.60	6.20	11.60
Standard stacking	86.64	71.38	3.09	15.82	10.27	12.8
Modified stacking	92.01	84.19	4.7	10.49	3.29	5.32

Table 8 Results of the experiment on the dataset obtained while monitoring cluster computational network

Number of tests	Number of correct solutions		Number of incorrect solutions			
	Training set	Test set	Errors of the 1st kind		Errors of the 2nd kind	
			Training set	Test set	Training set	Test set
Naive Bayes classifier	79.45	62.45	16.8	25.6	3.75	11.95
kNN classifier	82.8	67.74	12.8	20.2	4.4	12.06
SVM	80.13	66.47	12.02	19.93	7.85	13.6
Classification trees	81.23	69.3	11.07	17.23	7.7	13.47
Standard stacking	82.44	68.8	5.16	17.82	12.4	13.38
Modified stacking	92.30	85.46	4.7	10.49	3.00	4.05

Preparing data to use in machine learning comprises several steps. The first stage of the works believed to take about 60–70% of the time, that is cleaning, filtering, and transforming data into a format suitable for use in machine learning algorithms. The second stage deals with pre-processing and direct training of models.

Data pre-processing and cleaning are critical steps that enable the efficient use of a dataset for machine learning. Raw data are often corrupted, unreliable, and may contain missing values. Using such data in modelling can lead to wrong results. These tasks are a part of processing and analysing these groups and usually refer to the initial data mining that is used to determine and plan the necessary pre-processing.

A network monitoring system is necessary to monitor the state of the entire network infrastructure, with all devices and systems. Administrators can monitor all components of the network infrastructure, which uses a specific interface and exchanges information about its state using a standard protocol.

To identify various types of threats and failures, the monitoring system must track a large number of parameters of network components that are implemented at the channel, network, session, and application layers of the OSI model.

Observing these layers of the OSI model layers enables controlling the use of the system resources as well as finding faults linked to the equipment operation, which is necessary to maintain high reliability of the network infrastructure.

The technique of neural network use to solve tasks of detecting anomalous network activity is described in work [37]. The work also considers the technique for collecting data on the network activity and singling out parameters of network packets for further analysis. The author of the article monitors just one layer of the network infrastructure, which leads to reducing the capability to detect threats aimed at various objects of the network infrastructure.

To predict and classify anomalous traffic, the author of [39] uses a vector machine (SVM) taking into consideration the efficiency indices. The basic parameters include dispersion, autocorrelation and self-similarity. The advantage of this method is that it does not use packet headers. This enables implementing this method in real-time systems. The drawback of this approach is that the parameters that are transmitted include a small data set, due to which this method can efficiently and accurately detect anomalous traffic only for a short period of time.

The approach [8, 40] to network monitoring based on the technology of mobile and intelligent agents is studied in [8]. The drawback of this approach lies in the fact that data are not filtered from noise and non-informative parameters, which significantly complicates and sometimes even worsens the model of machine learning.

Before making the algorithm more complicated, it is necessary to make sure that the accuracy of its operation cannot be improved any longer by changing the parameters only. Therefore, one of the main parameters that requires careful optimization in each algorithm is the model complexity.

Complications must not start by the algorithm, but by adding parameters that have a physical interpretation that is significant for predicting the target value, domain experts usually formulate them. Since such features are often nonlinear transformations of the initial data, complex algorithms cannot reproduce them on their own. At the same time, since these parameters have a characteristic physical content, taking them into account in simple models often enables making these models more accurate than complex models that do not take these parameters into account.

The authors considered various techniques for pre-processing data and assessing the information content while determining the parameters of the network infrastructure control for more efficient data mining. The following tasks were formulated:

- to consider the methods of parameters selection;
- to determine a set of parameters to assess the state of the network using the data presented in [3] as an example.

Large-scale data mining is very important nowadays. Solving such a problem is often complicated because there are not very large sets or there are features uncorrelated in relation to the target variable, excessive features.

Good data is a prerequisite for creating good predicting models. To avoid the situation "garbage in, garbage out" and improve the quality of data and, as a consequence, the efficiency of the model, it is necessary to monitor the data health, identify problems as early as possible and decide what actions are necessary to pre-processed and clean the data [41–46]. Therefore, the task of selecting information features arises.

Based on the nature of the features, two main reasons for feature selection can be singled out:

- a large number of features, which significantly increases the operational time of the classifiers. Currently, ensemble machine learning methods are being developed and the time required for computation can become simply enormous through a large number of features. It can also result in service denial due to memory overflow. Because of this a hardware platform cannot be selected, which leads to the modification of classification algorithms for each platform separately.
- when the number of features increases, the prediction accuracy often decreases. Especially if the data contains a lot of noise features (that little correlate with the target variable). It also leads to the appearance of duplicated information signs, which results in overfitting.

The methods of feature selection can be grouped into three categories.

The *filtration methods* are organized by criteria that do not depend on the classification method, for example, feature correlation with the target vector, criteria of information content. This method is used before using classification algorithms. The advantage of filtering methods is that they can be used as pre-processing to reduce the dimensionality of a set of features and avoid overfitting.

Filters are used to select features in clustering, to build an initial approximation. The drawback of such methods is that they cannot identify complex relationships between features, and they are not good enough to identify all relationships in data.

An example of feature filtration is the mutual information method. This method is based on the concept of information entropy, and its formula is as follows:

$$H(X) = -\sum_{x_i \in X} p(x_i) * \log_2(p(x_i)), \tag{48}$$

where $p(x_i)$ is the probability of the fact that variable X gets x_i value. In this example, this probability is calculated as the number of examples in which $X = x_i$, divided by all examples.

To calculate the correlation between variables, two quantities are used:

$$H(Y|X = x_i), \tag{49}$$

the partial conditional entropy is the entropy $H(Y)$ calculated only for those records where $X = x_i$,

$$H(Y|X) = \sum_{x_i \in X} p(x_i) * H(Y|X = x_i), \tag{50}$$

the conditional entropy is the density of a continuous random variable distribution x.

The difference between these two values determines the degree of correlation (mutual information) between the values of X and Y, and shows how great it is.

$$IG(Y|X) = H(Y) - H(Y|X). \tag{51}$$

Wrapper methods are based on the information of the feature significance obtained with classification or regression methods and therefore can identify deeper patterns in the data than filters [47–49]. Wrappers can use any classifier that determines the degree of the feature significance.

There exist two approaches to the implementation of these methods—forward selection and backwards selection of features. The first ones start with an empty subset, to which various features are gradually added. In the second case, the method starts with a subset equal to the initial set of features from where features are gradually removed. In this case, the classifier is recalculated every time.

An example of such methods is the method of recursive feature elimination. As the name suggests, it refers to algorithms of the gradual exception of features from the general pool.

In the terms of the external assessment that assigns weight factors (for example, linear model coefficients) aimed at recursively eliminating features, the assessment algorithm first learns from the first set of features and determines the significance of each feature, and the least significant features are then removed from their current set. This procedure is repeated recursively until the desired number of features is eventually reached.

Built-in methods select features while the classifier is being trained, and optimize the set of features used to improve accuracy.

One of the main methods from this category is regularization. There are different implementations of this method, but the basic principle is the same. A classifier without regularization operates to build such a model that is best adjusted to predict all points in the training set.

For example, if the classification algorithm is a linear regression, then the polynomial coefficients are adjusted, which approximates the dependence between the features and the target variable.

The idea behind regularization is to build an algorithm that minimizes not only the error but the number of variables used.

The advantage of built-in algorithms is that they usually find a solution faster, avoiding re-training data from scratch and there is no need to separate the data into training and test sets. At the same time, no built-in methods are currently known to solve all existing problems.

An example of such methods is the method of Tikhonov regularization (or ridge regression) [50–52]. Consider it using the example of linear regression as well. If in a test set there is a matrix of features A and the vector of target variable b, the solution is $Ax = b$. In the process of algorithm operation, the expression is being minimized:

$$\|Ax - y\|^2 + \alpha \|x\|^2, \tag{52}$$

where the first addend is the root-mean-square error, and the second is the regularizing operator (the sum of the squares of all coefficients multiplied by the alpha). During the algorithm operation, the magnitudes of the coefficients will be proportional to the significance of the corresponding variables, and before the variables that eliminate the error least of all, they will tend to zero.

The alpha parameter enables adjusting the contribution of the regularizing operator to the total sum. With its help, the priority can be specified, that is the accuracy of the model or the minimum number of variables used.

To do comparative research, the dataset of the machine learning competition KDD 1999 was used as well as data obtained while monitoring the network infrastructure of the training data-center based the file system Lustre, that is detailed in Table 9. The data include 38 features, among which there 27 numerical features and 11 categorical ones.

Table 9 Network parameters

#	Parameters	#	Parameters
1	Connection duration, s	21	Number of connections to matching host
2	Transport layer protocol	22	Percentage of "SYN" error connection
3	Application layer service	23	Number of connections to one channel of the output port
4	Input stream, bytes	24	Percentage of "REJ" error connection
5	Output stream, bytes	25	Percentage of connections to the same service
6	Flags set in the TCP header	26	Percentage of connections to various services
9	Urgent data in the packet (URG flag)	27	Number of connections to the same service
10	Number of hot indicators	28	Percentage of source SYN error connections
11	Number of unsuccessful login attempts	31	Percentage of source REJ error connections
12	Successful login	32	Number of exports to MDT, including other Luster servers
13	Administrator access	33	Number of client connections by NID
14	Number of administrator access attempts	34	Number of locks
15	Number of Access Check File Operations	35	Luster-distributed lock manager (ldlm) has granted locks
16	Number of file creation operations	36	ldlm-lock of GR grant level
17	Number of Access Control File Operations	37	ldlm-lock of CP undo level
20	Guest system sign	38	Number of output commands in a ftp session

Table 10 Network parameters

Methods	Feature number
Info gain	5, 3, 6, 4, 30, 29, 33, 34, 35, 38,
Chi-squared	5, 3, 6, 4, 29, 30, 33, 34, 35, 12
ReliefF	3, 29, 4, 32, 38, 33, 30, 12, 36, 6
Variance threshold	3, 6, 4, 32, 29, 33, 30, 12, 36, 38

Fig. 25 Classification accuracy of filtering methods

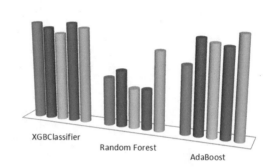

Table 10 presents features arranged in descending order of correlation values by the target variable.

The different ordering of features is due to the fact that each method uses own ranking algorithms.

To test the quality of classification, all data were divided into three parts to use three-folds cross-validation. XGBClassifier, Random Forest, AdaBoost were used as classifiers, and the algorithm error was assessed by MSE metric.

Figure 25 shows the mean error for three folds for each filtration method. As the chart shows, the quality of classification algorithm has not changes after filtration.

The time required to build a classification model, which is the duration of the classifier training process after applying each method, decreased significantly (Table 11).

At the next stage of the experiment a wrapper method, namely, Random Forest, was used. The number of trees in Random Forest varied from 100 to 500. Figure 26 presents a chart, which shows the significance of the first 12 features for three parts of the set using the algorithm built in Random Forest.

Other features had the significance less than 0,1. Based on the features that had maximum significance, a simple algorithm was built; its mean error by MSE metric is less different from XGBClassifier, but the speed of the algorithm operation increased.

Conclusion

The chapter proposes a technique to monitor the state of Self-healing Systems. As a result of using the technique, the indicators, used for detecting anomalies in the

Table 11 Efficincy index

Method	Algorithm	MSE (%)	Time (s)
Info gain	XGBClassifier	88.74	182
	Random Forest	84.34	167
	AdaBoost	86.12	170
Chi-squared	XGBClassifier	88.36	183
	Random Forest	85.08	163
	AdaBoost	88.49	172
ReliefF	XGBClassifier	87.92	183
	Random Forest	83.58	165
	AdaBoost	88.13	172
Variance threshold	XGBClassifier	88.92	182
	Random Forest	83.62	165
	AdaBoost	87.93	170
Full set	XGBClassifier	88.56	263
	Random Forest	87.12	223
	AdaBoost	89.07	243

Fig. 26 Significance of features

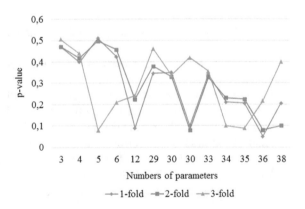

operation of Self-healing Systems, are improved by building models and methods based on data mining technologies.

The main results are as follows:

- The existing methods and means of monitoring Self-healing Systems were studied, which showed that there is no holistic assessment of the state of ShS operation.
- The features of the architecture of Self-healing Systems were considered, which enabled identifying the ShS basic components that require constant monitoring of their condition.
- A lot of parameters were determined to assess the state of each element of the system, which allows the operation of the entire system to be further assessed and the training time of the ensemble of classifiers be reduced.

- A technique and a set of models for the functioning of the multi-agent anomaly monitoring system during the operation of Self-healing Systems components were developed, based on the IDEF0 methodology, which detail the process of monitoring the ShS state and allow the monitoring system to be integrated with other system components.
- A method to detect anomalies in Self-healing Systems was developed based on stacking methods of data mining, which, in contrast to existing algorithms for detecting anomalies, can increase the detection accuracy up to 92% and reduce the number of the 2nd kind errors to 4% by using heterogeneous methods of data mining and up to the system training on the data obtained during the operation of the monitoring system.

References

1. Kora, A.D., Soidridine, M.M.: Nagios based enhanced IT management system. Int. J. Eng. Sci. Technol. (IJEST) **4**(4), 1199–1207 (2012)
2. Petruti, C.M., Puiu, B.A., Ivanciu, I.A., Dobrota, V.: Automatic management solution in cloud using NtopNG and Zabbix. In: 2018 17th RoEduNet Conference IEEE, Networking in Education and Research (RoEduNet), pp. 1–6 (2018)
3. Cigala, V.: Job-oriented monitoring of clusters. Int. J. Comput. Sci. Eng. **3**(3), 1333–1337 (2011). https://www.researchgate.net/publication/50418081_Job_Oriented_Monitoring_Clus ters
4. Stefanov, K.: Dynamically reconfigurable distributed modular monitoring system for super-computers (DiMMon). Procedia Comput. Sci. **66**, 625–634 (2015)
5. Sidorov, I., Sidorova, T., Kurzybova, Y.: Meta-monitoring system for ensuring a fault tolerance of the intelligent high-performance computing environment. In: ICCS-DE, pp. 99–107 (2019)
6. Tarasov, A.G.: Integration of computing cluster monitoring system. In: Proceedings of the First Russia and Pacific Conference on Computer Technology and Applications (RPC 2010), pp. 221–224 (2010)
7. Zaitseva, E., Levashenko, V.: Construction of a reliability structure function based on uncertain data. IEEE Trans. Reliab. **65**(4), 1710–1723 (2016)
8. Ruban, I., Martovitsky, V., Lukova-Chuiko, N.: Designing a monitoring model for cluster super-computers. Eastern-Eur. J. Enterp. Technol. **6**(2), 32–37 (2016)
9. Zaitseva, E., Levashenko, V.: Reliability analysis of multi-state system with application of multiple-valued logic. Int. J. Qual. Reliab. Manage. **34**(6), 862–878 (2017)
10. Svyrydov, A., Kuchuk, H., Tsiapa, O.: Improving efficiently of image recognition process: approach and case study. In: Proceedings of 2018 IEEE 9th International Conference on Dependable Systems, Services and Technologies, DESSERT 2018, pp. 593–597 (2018). https://doi.org/10.1109/DESSERT.2018.8409201
11. Zaitseva, E., Levashenko, V.: Multiple-valued logic mathematical approaches for multi-state system reliability analysis. J. Appl. Logic **11**(3), 350–362 (2013)
12. Kuchuk, G., Kovalenko, A., Komari, I.E., Svyrydov, A., Kharchenko, V.: Improving big data centers energy efficiency. Traffic based model and method. Stud. Syst. Decis. Control. **171**,161–183 (2019). https://doi.org/10.1007/978-3-030-00253-4_8
13. Xia, W., Liu, Y., Chen, D.: Construction of multitier distributed computing data mining system in cloud computing environment. In: 2017 2nd International Conference on Materials Science, Machinery and Energy Engineering (MSMEE 2017), pp. 1664–1667 (2017)

14. Gazafroudi, A.S., Pinto, T., Prieto-Castrillo, F., Prieto, J., Corchado, J.M., Jozi, A., Venayagamoorthy, G.K.: Organization-based multi-agent structure of the smart home electricity system. In: 2017 IEEE Congress on Evolutionary Computation (CEC), pp. 1327–1334 (2017)
15. Zhang, Y.H., Li, Z.T., Wang, M.Z., Xiao, L.: A multi-link aggregate IPSec model. In: 2009 First International Workshop on Education Technology and Computer Science, vol. 3, pp. 489–493 (2009). https://doi.org/10.1109/ETCS.2009.639
16. Mozhaev, O., Kuchuk, H., Kuchuk, N., Mykhailo, M., Lohvynenko, M.: Multiservice network security metric. In: 2nd International Conference on Advanced Information and Communication Technologies, AICT 2017, Proceedings, pp. 133–136 (2017). https://doi.org/10.1109/AIACT.2017.8020083
17. Merlac, V., Smatkov, S., Kuchuk, N., Nechausov, A.: Resourses distribution method of university e-learning on the hypercovergent platform. In: 2018 IEEE 9th International Conference on Dependable Systems, Service and Technologies, DESSERT'2018, Kyiv, pp. 136–140 (2018). https://doi.org/10.1109/DESSERT.2018.8409114
18. Ringberg, H., Soule, A., Rexford, J., Diot, C.: Sensitivity of PCA for traffic anomaly detection. In: Proceedings of the 2007 ACM SIGMETRICS International Conference on Measurement and Modeling of Computer Systems, pp. 109–120 (2007)
19. Kvassay, M., Levashenko, V., Zaitseva, E.: Analysis of minimal cut and path sets based on direct partial Boolean derivatives. Proc. Inst. Mechan. Eng. Part O: J. Risk Reliab. **230**(2), 147–161 (2016)
20. Gregg, B.: Systems Performance, Enterprise and the Cloud (2013)
21. Attar, H., Khosravi, M.R., Igorovich, S.S., Georgievan, K.N., Alhihi, M.: Review and performance evaluation of FIFO, PQ, CQ, FQ, and WFQ algorithms in multimedia wireless sensor networks. Int. J. Distrib. Sens. Netw. **16**(6), 155014772091323 (2020). https://doi.org/10.1177/1550147720913233
22. Ruban, I.V., Martovytskyi, V.O., Kovalenko, A.A., Lukova-Chuiko, N.V.: Identification in informative systems on the basis of users' behaviour. In: Proceedings of the International Conference on Advanced Optoelectronics and Lasers, CAOL 2019-September, vol. 9019446, pp. 574–577 (2019). https://doi.org/10.1109/CAOL46282.2019.9019446
23. Mukhin, V., Kuchuk, N., Kosenko, N., Kuchuk, H., Kosenko, V.: Decomposition method for synthesizing the computer system architecture. Adv. Intel. Syst. Comput. AISC. **938**, 289–300 (2020). https://doi.org/10.1007/978-3-030-16621-2_27
24. Ekanayake, J., Fox, G.: High performance parallel computing with clouds and cloud technologies. In: International Conference on Cloud Computing. Springer, Berlin, Heidelberg, pp. 20–38 (2009)
25. Kharchenko, V., Kovalenko, A., Andrashov, A., Siora, A.: Cyber security of FPGA-based NPP I&C systems. Challenges and solutions. In: 8th International Topical Meeting on Nuclear Plant Instrumentation, Control, and Human-Machine Interface Technologies 2012, NPIC and HMIT 2012: Enabling the Future of Nuclear Energy, 2012, vol. 2, pp. 1338–1349 (2012)
26. Kovalenko, A., Kuchuk, H., Kuchuk, N., Kostolny, J.: Horizontal scaling method for a hyperconverged network. In: 2021 Int. Conference on Information and Digital Technologies (IDT), Zilina, Slovakia (2021). https://doi.org/10.1109/IDT52577.2021.9497534
27. Levashenko, V., Zaitseva, E., Kvassay, M., Deserno, T.M.: Reliability estimation of healthcare systems using Fuzzy Decision Trees. In: Proceedings of the Federated Conference on Computer Science and Information Systems (FedCSIS), Gdansk, Poland, Sept. 11–14, pp. 331–340 (2016)
28. Stallings, W.: SNMP, SNMPv2, SNMPv3, and RMON 1 and 2. Addison-Wesley Longman Publishing Co., Inc. (1998)
29. Yakimov, I.M., Trusfus, M.V., Mokshin, V.V., Kirpichnikov, A.P.: AnyLogic, extendsim and simulink overview comparison of structural and simulation modelling systems. In: 2018 3rd Russian-Pacific Conference on Computer Technology and Applications (RPC) (pp. 1–5). IEEE (2018, August)
30. Onan, A.: Classifier and feature set ensembles for web page classification. J. Inform. Sci. **42**, 150–165 (2016)

31. Baskin, I.I.: Bagging and boosting of classification models. Tutorials Chemoinform. **15**, 241–247 (2017). https://doi.org/10.1002/9781119161110.ch15
32. Kuchuk, G.A., Akimova, Yu.A., Klimenko, L.A.: Method of optimal allocation of relational tables. Eng. Simul. **17**(5), 681–689 (2000)
33. Liu, H.: Comparison of four Adaboost algorithm based artificial neural networks in wind speed predictions. Energy Convers. Manage. **92**, 67–81 (2015)
34. Levashenko, V., Lukyanchuk, I., Zaitseva, E., Kvassay, M., Rabcan, J., Rusnak, P.: Development of programmable logic array for multiple-valued logic functions. IEEE Trans. Comput.-Aided Design Integr. Circuit. Syst. **39**(12), 4854–4866 (2020)
35. Wolpert, D.H.: Stacked generalization. Neural Netw. **5**, 241–259 (1992)
36. Niu, Z.: 2d cascaded adaboost for eye localization. In: 18th International Conference on Pattern Recognition (ICPR'06). **2**, 1216–1219 (2006)
37. Joshi, S., Sherali, H., Tew, J.: An enhanced response surface methodology (RSM) algorithm using gradient deflection and second-order search strategies. Comput. Operat. Res. **25**, 531–541 (1998)
38. Semenov, S., Sira, O., Gavrylenko, S., Kuchuk, N.: Identification of the state of an object under conditions of fuzzy input data. East.-Eur. J. Enterp. Technol. **1**(4), 22–30 (2019). https://doi.org/10.15587/1729-4061.2019.157085
39. Kuchuk, N., Shefer O., Cherneva G., Ali, A.F.: Determining the capacity of the self-healing network segment. Adv. Inform. Syst. **5**(2), 114–119 (2021). https://doi.org/10.20998/2522-9052.2021.2.16
40. Martovytskyi, V., Ruban, I., Lukova-Chuiko, N.: Approach to classifying the state of a network based on statistical parameters for detecting anomalies in the information structure of a computing system. Cybern. Syst. Anal. **54**, 302–309 (2018)
41. Zinchenko O., Vyshnivskyi V., Berezovska Yu., Sedlaček P.: Efficiency of computer networks with SDN in the conditions of incomplete information on reliability. Adv. Inform. Syst. **5**(2), 103–107 (2021). https://doi.org/10.20998/2522-9052.2021.2.14
42. Nykolaichuk, Y., Pitukh, I., Vozna, N., Protsiuk, H., Nykolaichuk, L., Volynskyy, O.: System for monitoring the quasi-stationary technological processes based on image-cluster model: In: 2017 9th IEEE International Conference on Intelligent Data Acquisition and Advanced Computing Systems: Technology and Applications (IDAACS, vol. 2, 712–715 (2017)
43. Zaitseva, E., Levashenko, V., Lukyanchuk, I., Rabcan, J., Kvassay, M., Rusnak, P.: Application of generalized Reed–Muller expression for development of non-binary circuits. Electronics (Switzerland) **9**(1), (2020). Article number 12(4)
44. Donets, V., Kuchuk, N., Shmatkov, S.: Development of software of e-learning information system synthesis modeling process. Adv. Inform. Syst. **2**(2), 117–121 (2018). https://doi.org/10.20998/2522-9052.2018.2.20
45. Rabcan, J., Levashenko, V., Zaitseva, E., Kvassay, M.: Review of methods for EEG signal classification and development of new fuzzy classification-based approach. IEEE Access **8**, 189720–189734 (2020)
46. Rabcan, J., Levashenko, V., Zaitseva, E., Kvassay, M.: EEG signal classification based on fuzzy classifiers. IEEE Trans. Indus. Inform. **18**, 757–766 (2021)
47. Kuchuk, H., Kovalenko, A., Ibrahim, B.F., Ruban, I.: Adaptive compression method for video information. Int. J. Adv. Trends Comput. Sci. Eng. 66–69 (2019). https://doi.org/10.30534/ijatcse/2019/1181.22019
48. Kovalenko, A., Kuchuk, H.: Methods for synthesis of informational and technical structures of critical application object's control system. Adv. Inform. Syst. **2**(1), 22–27 (2018). https://doi.org/10.20998/2522-9052.2018.1.04
49. Tibshirani, R.J.: Exact post-selection inference for sequential regression procedures. J. Am. Stat. Assoc. **111**(514), 600–620 (2016)
50. Sobchuk V., Zamrii I., Olimpiyeva Yu., Laptiev S.: Functional stability of technological processes based on nonlinear dynamics with the application of neural networks. Adv. Inform. Syst. **5**(2), 49–57 (2021). https://doi.org/10.20998/2522-9052.2021.2.08

51. Frei, R., McWilliam, R., Derrick, B., Purvis, A., Tiwari, A., Serugendo, G.D.M.: Self-healing and self-repairing technologies. Int. J. Adv. Manuf. Technol. **69**, 1033–1061 (2013). https://doi.org/10.1007/s00170-013-5070-2
52. Ghosh, D., Sharman, R., Rao, H.R. Upadhyaya, S.: Self-healing systems—survey and synthesis. Decis. Support Syst. **42**(4), 2164–2185 (2007). https://doi.org/10.1016/j.dss.2006.06.011

Self-healing Systems Modelling

Nina Kuchuk and Vitalii Tkachov

Abstract The chapter proposes a set of Self-healing Systems models. The proposed models make it possible to take into account the features of Self-healing Systems, in particular, the fact that there exists a mechanism to monitor and detect failures and the auto-recovery protocols. Also, the models enable planning approaches to improve QoS parameters and reduce the cost to operate the core network of Self-healing Systems. The general principles of building the structures of Self-healing Systems are considered. A mathematical model of the information structure of Self-healing Systems is proposed. A fast algorithm is developed to determine whether the current solution belongs to the space of feasible solutions. The developed model takes into account the features of Self-healing Systems and enables taking into account informational relationships between its components. A model of the technical structure of the core network of Self-healing Systems is proposed. A feature of this model is that it is isomorphic to the information model and is built on its basis. The peculiarities of information transmission in wireless components of Self-healing Systems are considered. Traffic models of wireless components of Self-healing Systems, focused on the features of such systems, are developed. The article illustrates that the use of the proposed modifications makes it possible to increase the bandwidth of the TCP in wireless components due to the reallocation of the bandwidth of the core network component. The chapter also formulates the problem that arises while building the models of Self-healing Systems topological structures, which take into account the evolution dynamics of specific components of the core network. A mathematical model that takes into account the evolution dynamics of specific network components, depending on their type and purpose is proposed.

Keywords Self-healing systems · Core network · Information structure ·
Technical structure · System evolution

N. Kuchuk (✉)
National Technical University «Kharkiv Polytechnic Institute», 2, Kyrpychova str., Kharkiv 61166, Ukraine
e-mail: nina_kuchuk@ukr.net

V. Tkachov
Kharkiv National University of Radio Electronics, 14 Nauki ave., Kharkiv 61166, Ukraine
e-mail: vitalii.tkachov@nure.ua

1 General Principles of Building the Structure of Self-healing System

Self-healing Systems (ShS) differ from other systems because they have both the mechanism of failure monitoring and detecting and the in-built protocol of auto-recovery [1, 2]. Therefore, ShS require additional Hardware and Software. This leads to changes in the system structure and requires relevant models that take into account the given features.

The ShS structure is the basic factor that affects the quality of data exchange between system applications and, consequently, on the quality of the solution of applied tasks a computer system makes [3, 4]. So, the analysis of the structure is a prerequisite while selecting options to build and manage the basic ShS network.

The main purpose of the structure analysis is to determine the parameters of data streams that pass through the communication channels of the core network (CN) of ShS and enter its nodes. These data enable assessing the load of communication channels and CN equipment. However, data streams cannot be studied only by means of the tasks of the network structure in their classical understanding, as a set of nodes and connections between them [5]. This is due to the fact that they are formed by CN applications that run on its nodes and exchange data with each other. So, to analyse the CN, it is necessary to supplement information about the structure with information about the application, their interaction and location on the nodes.

The results of the analysis must be the numerical values of the CN throughput: the load on the communication channels and structural equipment, the rate of data streams and requests entering the network nodes. In this case, these characteristics must be calculated taking into account the peculiarities of a specific CN structure.

The problems of analysing the ShS structure are as follows [6, 7].

First, there is no unified approach to build the structure.

Second, there is an explicit dependence of the characteristics of the ShS structure on the parameters of applied problems that are solved in the CN environment, for example, the number of applications and their assigning to the network nodes.

Third, there are no proven mathematical methods for the formal description of the ShS structure that could be used in calculations. All this requires that general principles to analyse the ShS structure, be developed, which are invariant with respect to:

- networking technologies;
- applied problems solved in the CN;
- equipment used to build the CN.

Quite often, to calculate the parameters of data streams as well as the load of CN nodes, mathematical models in the form of queuing networks (QS, Queuing System) are used [8]. Using QS as a ShS model enables analysing the operation of a system with a complex structure and various disciplines for query serving that run various applications. However, the use of QS for ShS analysis is related to such features [9, 10]:

- data entering the CN of ShS through communication channels are considered in QS as requests for serving, which is not exactly correct since part of the transmitted data is just information necessary for the operation of applications;
- when analysing the system using QS, it is assumed that the rate of requests that enter the node is equal to the rate of requests that leave the node, but it is not always true for CN of ShS and depends on the application features;
- in CN of ShS, the load of each network communication channel is determined by the rate of data streams of all types transmitted over the channel but not only by the stream of requests;
- in ShS, resources are allocated taking into account the capacity of operational self-healing of the system, which complicates the analysis using QS;
- ShS uses communication channels and network nodes to transmit of transient data streams and requests for serving, which are not taken into account in QS;
- many results of the QS analysis, such as delays in queues of requests for serving, are not necessary for the administrator of the CN of ShS, and their calculations are connected with significant resource and time consumption.

The fact that there are specified features of the CN of ShS operation often leads to the need to neglect some of them while using QS as models, which results in a loss in the accuracy of the modelling results. In addition, a rather complex CN analysis is required when preparing data for modelling, which makes it difficult to use QS to solve general problems of ShS analysis. Another factor constraining the use of QS for ShS analysis is their large dimension. Actual networks have thousands of nodes, and this makes it almost impossible to apply the methods of describing the structure and distribution of streams developed for QS [11].

In this case, the basic principles to build the ShS structure are as follows:

- the main purpose of the analysis is to study data streams in CN, which are the main factors affecting all ShS characteristics;
- the basis for the analysis and formation of the CN structure is the executable and interacting applications;
- when analysing, it is necessary to match the requirements for the operation of applications with the capabilities of the network equipment.

Therefore, it is reasonable to apply an approach to analysing and building the network structure based on the study of the interaction of applications and tasks as independent sources and receivers of data in CN. In this case, it is possible to assign the parameters of data streams between applications when performing the entire complex of tasks (to build the ShS information model), and then, depending on the assigning of applications to network nodes, to assign the parameters of data streams between CN nodes (to build a technical CN model). In this case, all interactions between applications are fully taken into account. Another advantage of this approach is the capability to analyse complex network structures by decomposing them into subnets, which is used in VLAN and VPN technologies [12].

To analyse the structure and calculate the CN characteristics according to the proposed approach, it is necessary to determine the rules of its description that enable building models to calculate the load of nodes and network communication channels.

By means of transactions, ShS users initiate the operation of applications, which, in turn, refer to databases and data stores, to other applications, transmitting and receiving various information. The operation of applications is determined by the tasks that the system solves.

An application is considered as a program that is run by a user while solving a problem; the program can be either special, written to solve a problem, or a system program designed to perform standard procedures that are also needed when solving a problem.

Each ShS has databases or data stores, and the applications that work with them are determined.

The main factors that affect the decisions made while building a CN are the tasks that are solved in the system environment. Therefore, to build a network, it is necessary to know the ShS information structure, which allocates the information streams between the nodes on which the software is installed.

A CN information structure is considered as a set of information resources ShS (sources and receivers of information) assigned to the network nodes and information streams between nodes that arise when solving problems. A node of the information structure means the resource (application, database) point where its operation is ensured. Data on the network information structure enable making a decision on the organization of communication channels between the network nodes, determining the necessary parameters of communication channels and network equipment, building the network physical structure.

A CN technical structure is understood as a set of network equipment, network nodes, and communication channels that constitute a fully connected network (a network where data transfer between any nodes is possible). A technical structure node is a set of technical means that implement the node of the information structure and ensure the operation of the installed resources. In this case, the node of the technical structure is a rather complex system, including several computers connected to a local network.

Thus, for an effective analysis of the ShS structure, the components of its information and technical structures must be analysed, and connect the analysis results.

This is due to the fact that the information structure determines the structure and parameters of data streams between applications and nodes (information nodes), and the technical structure, using the results of the information structure analysis, determines specific routes of data transmission and network characteristics, ways of implementing information structure nodes and building network nodes to form a technical structure.

Connecting the results of the analysis of information and technical structures involves mapping the characteristics of the ShS information structure to the characteristics of the CN technical structure and determining the parameters of the technical structure based on the parameters and characteristics of the information structure.

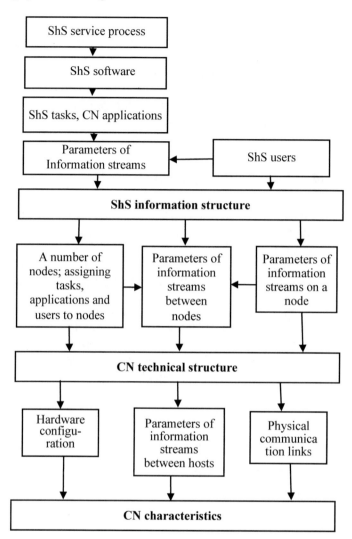

Fig. 1 Analysis chart of the ShS structure

The analysis chart of the ShS structure is shown in Fig. 1.

2 Modelling the Information System of Self-healing Systems

Consider the process of developing a model of the ShS information structure step-wise. At the first stage of development, a formalized description of the information structure of ShS is carried out. At the second stage, a set of feasible options for the information structure of the system is determined, on the basis of which the optimal option for the technical structure of the core network can be selected. At the third, final, stage, the optimal information structure of the system is chosen according to the selected criterion of the quality of ShS operation. To do this, a method for constructing and solving the appropriate optimization problem is developed using the mathematical apparatus of genetic algorithms.

2.1 Formalized Description of the Information Structure

The basic controlling node, that is responsible for the coordination of all ShS tasks in CN, is a CN hypervisor. It accepts the transactions of the system users and runs the appropriate system and user applications [13]. In turn, applications can be conditionally classified into two groups: those that directly operate with the system information nodes (ZA group) and those that directly make exchange with the distributed data store (ZB group).

Assume, for simplicity, that every application, no matter what task it is used for, always operates the same way [14, 15]. So, if the operation of the application changes depending on the task, this application is understood as another one.

Let us identify the main sets that are involved in the process of modelling information relationships in the ShS environment:

M_U is a set of ShS users, dim $M_U = U$;

M_N is a set of CN information nodes, dim $M_N = N$;

M_A is a set of CN applications, dim $M_A = A$;

M_E is a set of ShS transactions, dim $M_E = E$;

M_D is a set of data store chunks, dim $M_D = D$.

A generalized diagram of the main information relationships in the environment of a computer system is given in Fig. 2.

The parameters of each transaction $e \in M_E$ ($e = \overline{1, E}$) is given by the following tuple:

$$E_e = \langle A_e, D_e, U_e, W_e \rangle, \tag{1}$$

where $A_e = (a_{ea}) = (a_{e1}, \ldots, a_{eA})$ is a vector of ShS application that are necessary for e transaction;

$D_e = (d_{ed}) = (d_{e1}, \ldots, d_{eD})$ is a vector of necessary ShS data store chunks;

$U_e = (u_{eu}) = (u_{e1}, \ldots, u_{eU})$ is a vector of users that run a transaction;

$W_e = (w_{ij})$ is a matrix of matrix of transaction application run e, $i, j = \overline{1, A}$.

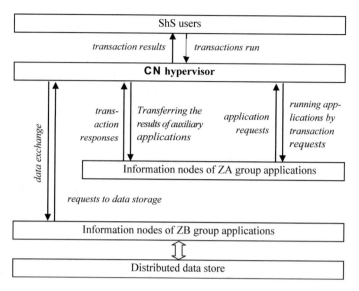

Fig. 2 Generalized diagram of the main information relationships in the ShS environment

All above vectors are Boolean vectors and take a nonzero value if and only if the appropriate condition is met, that is, the transaction e is used or uses the appropriate resource. Matrix $\overline{W_e}$ is also a Boolean matrix where element $w_{ij} = 1$ if and only if application j can be performed after application i is completely finished.

The parameters of application a of e transaction are given by a tuple

$$A_{ae} = \langle \overline{\lambda_{ae}}, \overline{\beta_{ae}} \rangle, \tag{2}$$

where $\overline{\lambda}_{ae} = (\lambda_{ae1}, \ldots, \lambda_{aeD})$ is data amounts that a application of e transaction must obtain and which are in an appropriate data store chunk; $\beta_{ae} = (\beta_{ae1}, \ldots, \beta_{aeA})$ is data amounts to exchange with other applications while completing e transaction.

Assigning ShS applications to CN nodes is given by Boolean matrix

$$G = (g_{an}), \tag{3}$$

where $g_{an} = 1$ if and only if application numbered a is run from the information node n, $a \in \overline{1, A}$, $n \in \overline{1, N}$, that is, rectangular matrix G is of $A \times N$ size.

Assigning users to CN nodes is given by Boolean matrix

$$H = (h_{un}). \tag{4}$$

where $h_{un} = 1$ if and only if n information node is assigned to application numbered u, and $u \in \overline{1, U}$, $n \in \overline{1, N}$, that is rectangular matrix H is of $U \times N$ size.

Assigning data store chunks to CN nodes is determined by the following matrix:

$$S = (s_{dn}).$$ (5)

where $s_{dn} = 1$ is and only if a data store chunk numbered d is assigned to n information node, $d \in \overline{1, D}$, $n \in \overline{1, N}$, that is S rectangular matrix is of $D \times N$ size.

Thus, a model can be formed by the following tuple of sets and matrices:

$$\aleph = \langle M_U, M_N, M_A, M_E, M_D, \{E_e\}, \{A_{ae}\}, G, H, S \rangle.$$ (6)

In this case, such constraints on the elements of the given tuple must be taken into account (6):

- any transaction uses at least one application, that is

$$\sum_{i=1}^{A} a_{ei} \geq 1 \quad \forall e \in \overline{1, E};$$ (7)

- the transaction must not necessarily access the data store, that is

$$\sum_{i=1}^{D} d_{ei} \geq 1 \quad \forall e \in \overline{1, E};$$ (8)

- each transaction is necessary for at least one user:

$$\sum_{i=1}^{U} u_{ei} \geq 1 \quad \forall e \in \overline{1, E};$$ (9)

- every active user must run at least one transaction, that is:

$$\sum_{i=1}^{E} u_{iu} \geq 1 \quad \forall u \in \overline{1, U};$$ (10)

- each application must be assigned to just one node:

$$\sum_{i=1}^{N} g_{ai} = 1 \quad \forall a \in \overline{1, A};$$ (11)

- each node has at least one active application (users are assigned to just information nodes):

$$\sum_{i=1}^{A} g_{in} \geq 1 \quad \forall n \in \overline{1, N};$$ (12)

- all applications must be installed, that is:

$$\sum_{i=1}^{A} \sum_{j=1}^{N} g_{ij} = A;$$ (13)

- all users must be assigned to information nodes, that is:

$$\sum_{i=1}^{U} \sum_{j=1}^{N} h_{ij} = N; \qquad (14)$$

- each user must be assigned to just one information node, that is:

$$\sum_{i=1}^{N} h_{ui} = 1 \quad \forall u \in \overline{1, U}; \qquad (15)$$

- assigning users to information nodes is arbitrary, that is:

$$\sum_{i=1}^{U} h_{in} \in [0, U]; \qquad (16)$$

- all data store chunks are assigned to the network nodes, that is:

$$\sum_{i=1}^{U} \sum_{j=1}^{N} s_{ij} = D; \qquad (17)$$

- there can exist data replication, that is, one chunk can be placed on several nodes:

$$\sum_{i=1}^{N} s_{di} \geq 1; \qquad (18)$$

- assigning data store chunks to ShS core network is arbitrary, that is:

$$\sum_{i=1}^{D} s_{in} \in [0, D] \quad \forall i \in \overline{1, N}. \qquad (19)$$

Therefore, tuple (6) along with conditions (7)–(19) make up a mathematical model of the ShS information structure. The model takes into account the features of this platform and enables establishing information relationships between ShS components and conduct core network analysis.

But in the developed model, it is necessary to determine the set of feasible options for the information structure of the system based on which the optimal technical structure of the core network can be selected.

2.2 Determining Feasible Options to Build Information Structure of Self-healing Systems

In tuple (19), matrices G, H, S that specify the assignment of information structure elements to its nodes are variables given in expressions (3)–(5). But when selecting the desired solution, it is necessary to check it for constraints (6)–(19), that is, to check if this solution is included in the space of feasible decision. Feasible solutions in three-dimensional space of rectangular Boolean matrices can be found in many ways. But, despite the fact that the time to determine the space of feasible solutions is constrained in many algorithms for finding an optimal or rational solution, a fast algorithm for determining the membership of the current solution in a given space is proposed.

Note, that rectangular matrices G, H, S have the same number of columns, so the concatenation by columns can be conducted, that is, a new combined matrix can be formed that specifies a variant of the information structure using the operation of the vertical concatenation of matrices:

$$G_{common} = G !! H !! S, \quad card(G_{common}) = (A + U + D) \times N. \quad (20)$$

Let us analyse the constraints in the mathematical model.

Since the transaction of users are run from the same information node where the appropriate user software is installed, the constraints on transactions (7)–(10) do not affect G_{common} matrix.

Taking into account the fact that G_{common} matrix is Boolean, constraints in assigning users and applications just to one information node (according to 15) are checked fast if the following equality is confirmed:

$$\sum_{i=1}^{A+U} \left(\varsigma_1 \left(\sum_{j=1}^{N} G_{common}(i, j) \right) \right) = A + U, \quad (21)$$

where a function for determining exactly one unit in a row or column of a Boolean matrix is used

$$\varsigma_1(\tau) = \begin{cases} 1, & if \ \tau = 1; \\ 0, & if \ \tau \neq 1. \end{cases} \quad (22)$$

Meeting condition (21) also corresponds to holding constraints on the obligatory assignment of all users and available applications (13) and (14) to information nodes.

To check constraint (12) as for the fact that there is at least one active application at each information node, which is a prerequisite for the core network of ShS, the chunks of the appropriate columns of the combined matrix are analyzed if the following equality is confirmed:

$$\sum_{j=1}^{N} \left(\zeta_2 \left(\sum_{i=1}^{A} G_{common}(i, j) \right) \right) = N, \quad (23)$$

where a function for determining non-zero chunk in a row or column of a Boolean matrix is used

$$\zeta_2(\tau) = \begin{cases} 1, & if\ \tau \ge 1; \\ 0, & otherwise. \end{cases} \tag{24}$$

Using function (24) also enables checking constraints as for obligatory placement of all data store chunk at the information nodes:

$$\sum_{j=1}^{N} \left(\zeta_3 \left(\sum_{i=A+U+1}^{A+U+D} G_{common}(i,\ j) \right) \right) = N. \tag{25}$$

Meeting condition (25) does not contradict the capability of the solution being considered regarding the data replication and arbitrary assignment of data store chunks at the information nodes of ShS (18) and (19).

Thus, the fast algorithm to determine whether the current solution belongs to the area of feasible solutions is as follows:

1. Vertical concatenation of basic matrices for solving and obtaining a generalizing matrix.
2. Checking whether there is exactly one unit in the rows that characterize applications and users.
3. Checking whether there is at least one active application at each information node.
4. Checking the constraints related to the location of data store chunks.

Thus, the developed mathematical model of the ShS information structure and the proposed fast algorithm to determine whether the current solution belongs to the given space enable developing a method for finding the optimal solution according to the selected criterion of the quality of ShS operation and will be considered at the next stage.

2.3   Statement of a Problem to Find Optimal Information Structure of Self-healing Systems

The proposed mathematical model of the ShS information structure has Boolean matrices of assigning applications, data store chunks, and ShS users to CN nodes as variable input data. Consequently, if the criteria of the developed model quality are given, an optimal or close to the optimal option of its building can be found by changing matrices (3)–(5).

Taking into account the ShS features described in the previous section, let us choose the balance of the load on its nodes as a criterion for the developed structure quality that is located on the ShS core network, and, accordingly, the sum of deviations of the node load from the average load as the indicator, that is, the less this indicator is, the higher the quality indicators of the computer system operation are.

The proposed criterion essentially depends on a load of communication channels between the nodes of the ShS core network, that is, on the parameters of information streams distributed over the selected information structure.

Consider information structure $\Im$ that is unambiguously determined by matrices (2)–(4). To make calculations, it is necessary to know the matrix of the rate of transactions users run

$$\left(\Omega = \left(\omega_{i,j}\right); \quad \omega_{i,j} \geq 0; \quad i = 1, \ldots, U; \quad j = 1, \ldots E\right). \tag{26}$$

Then, the total rate of requests to run the j-th transaction ω_j can be calculated as well as the vector of transactions activity $\overline{\omega}$:

$$\overline{\omega} = \left(\omega_j\right). \tag{27}$$

Using expressions (1) та (27), it is possible to calculate the total rate of running the k-th ($k = 1, \ldots, A$) application θ_k by all the system transactions as well as the vector of the rate of running applications θ, where $\alpha_{j,k}$ is a Boolean vector of applications necessary for the j-th transaction:

$$\theta_k = \sum_{j=1}^{E} \omega_j \alpha_{j,k}, \quad \theta = (\theta_k). \tag{28}$$

Let us consider how the j-th transaction operates at the information nodes n and m ($n, m = 1, \ldots, N$). First, the load of m node is calculated by requests of the j-th transaction to use application ξ by expressions (26)–(28):

$$Z_{jm\xi 1} = \sum_{\eta=1}^{D} \alpha_{j,\eta} g_{\eta,m} \beta_{\xi,\eta}.$$

Then the load of n node with the requests of the j-th transaction that forms the m node is as follows:

$$Z_{jnm1} = \sum_{\xi=1}^{A} \alpha_{j,\xi} g_{\xi,n} Z_{jm\xi 1}. \tag{29}$$

The amounts of data that are transferred from data stores are calculated similarly while making requests of the j-th transaction, loading information nodes n and m:

$$Z_{jnm2} = \sum_{\xi=1}^{A} \alpha_{j,\xi} g_{\xi,n} \left(\sum_{\eta=1}^{D} d_{j,\eta} s_{\eta,m} \lambda_{\xi,\eta}\right). \tag{30}$$

Proceeding from Eqs. (28) and (30), the amounts of data that operate in the network between information nodes n and m can be calculated, meeting the requirements of the j-th transaction $Z_{n,m}$, as well as the matrix of loading the channels of core network of ShS Z_j by this transaction:

$$Z_{j,n,m} = \sum_{j=1}^{E} (Z_{jnm1} + Z_{jnm2}), \quad Z_j = (Z_{j,n,m}). \tag{31}$$

This expression enables building a matrix of the exchange rate between network information nodes while making the j-th transaction:

$$C_j = (C_{j,n,m}), \quad C_{j,n,m} = \omega_j Z_{j,n,m}. \tag{32}$$

Consequently, the matrix of the exchange rate between information nodes (this is also the load of communication channels and network equipment) is determined by the following formula:

$$C = (C_{n,m}), \quad C_{n,m} = \sum_{j=1}^{A} C_{j,n,m} \tag{33}$$

This makes it possible to determine the average load of one information node of the network:

$$C_{average} = \left(\sum_{n=1}^{N} \sum_{m=1}^{N} C_{n,m} \right) / N. \tag{34}$$

Taking into account the fact that the selected criterion to characterize the target structure is the balance of load at the nodes, and the indicator is the sum of load deviations from the average one, the value of the indicator with a fixed information structure $\Im$ can be determined:

$$\Psi(\Im) = \Psi(G, H, S) = \sum_{n=1}^{N} \sum_{m=1}^{N} |C_{n,m} - C_{average}|. \tag{35}$$

Therefore, the target function of the task to find the optimal ShS information structure that is being studied is as follows:

$$\Psi(\Im) \xrightarrow{\Im} \min. \tag{36}$$

Also, using the obtained ratios, the load of individual information network nodes can be calculated. The rate of request stream to run application i, installed at n node, is as follows:

$$b_{i,n} = \sum_{j=1}^{A} \theta_i \cdot g_{j,n} \tag{37}$$

The obtained values are non-zero only in the case if application i is installed at n node, they enable building the matrix of application rate at the network nodes

$$B = (b_{i,n}). \tag{38}$$

The value of n node load with the elements of data store is formed: the j-th transaction, the n-th node, the d-th data block:

$$\varphi_{j,n,d} = \sum_{i=1}^{A} \sum_{k=1}^{D} \left(\theta_j \cdot s_{k,n} \cdot d_{j,d} \cdot g_{i,n} \right). \tag{39}$$

Then, the rate of requests to d data block at n node is equal to:

$$\varphi_{n,d} = \sum_{j=1}^{E} \varphi_{j,n,d}, \tag{40}$$

and the rate of requests to the data storea is determined by matrix $\varphi = \left(\varphi_{n,d} \right)$.

Thus, to find the optimal ShS information structure, it is necessary to solve a task of nonlinear discrete programming with an objective function (36) that is formulated by ratios (27)–(35) and constraints (7)–(19). Solving this task by the full enumeration method requires analyzing options

$$\aleph = N^A \cdot N^U \cdot N^D = N^{(A+U+D)}.$$

This number of options is unacceptably large even for small computer networks. Other methods of finding the exact solution are also costly. Therefore, it is proposed to use approximate methods that enable obtaining a solution close to optimal, or rational.

2.4 Quick Search for a Rational Structure Using a Genetic Algorithm

From the approximate methods, taking into account a very large and difficult search space, as well as a rather simple algorithm for calculating the selected structure quality indicator, a genetic algorithm was chosen to find an approximate solution [16].

Consider the steps of setting up the appropriate algorithm.

The search space consists of three clusters that are given by Boolean matrices (2)–(4). These clusters form a population that has $A + U + D$ individuals. Each individual is determined exactly by one chromosome $\zeta = (\zeta_i)$. Each chromosome, taking into account constraints (11), (15) and (17), has $3 \cdot N$ genes (that is the size of appropriate vector card $\zeta = 3 \cdot N$), that are formed from the three current clusters of the search space as follows:

- all three parts have the same length—N genes;
- the genes of the first part are formed from matrix $H = (h_{un})$ by summing up its columns, and it must be noted that constraint (14) provides the Boolean character of the sum, and 1 corresponds to assigning a user to the appropriate information node;

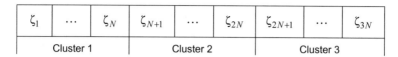

ζ_1	$\cdots$	ζ_N	ζ_{N+1}	$\cdots$	ζ_{2N}	ζ_{2N+1}	$\cdots$	ζ_{3N}
Cluster 1			Cluster 2			Cluster 3		

Fig. 3 Chromosome structure

- the genes of the second part are formed from matrix $S = (s_{dn})$ by summing up its columns and constraint (17) provides the Boolean character of the sum, and 1 corresponds to assigning a data store chunk to the appropriate information node;
- the genes of the third part are formed from matrix $G = (g_{an})$ by summing up its columns, because several applications can be assigned to one information node, but according to constrain (11) the Boolean character of the sum is provided, and 1 corresponds to assigning an appropriate application to a particular information node.

The chromosome structure is given in Fig. 3.

When forming individuals and while using a crossover, it is checked whether the chromosome of the obtained individual belongs to the search space. Each chromosome corresponds to the assignment of users, applications, and data store chunks to information nodes, therefore it is rather simple to determine the fitness function that characterizes the chromosome quality, and, consequently, a specific individual according to the selected indicator of the balance of the load on the nodes, which is preliminarily grounded and given by expression (35) and is compared using the criterion (36). Thus, for each population there is a fixed average load of one information node of the network, that is, the fitness function of each individual is the deviation value of the appropriate information node, calculated according to expression (36) from the average value, and the greater this value is, the less the individual is used in this population.

Therefore, under the given conditions an appropriate genetic algorithm can be proposed, which enables finding a rational solution of the proposed method in a reasonable period of time.

Step 1. Creation of an initial population. Before the first step, it is necessary that an initial population with a predetermined number of individuals be created at random. Even if the population turns out to be completely uncompetitive, the genetic algorithm quickly converts it into a population appropriate for further development. Thus, at the first stage, it is sufficient that all its elements (in the form of chromosomes) correspond to the format of the population individuals, and the fitness function can be calculated for each chromosome. The consequence of the first step is a population of $A+U+D$ individuals.

Step 2. Calculations of the fitness function for the population individuals. The quality indicator is used as the fitness function of the entire population, it is calculated for all individuals of the population according to formula (35), based on which the fitness functions of separate individuals are calculated.

Step 3.   Selection of individuals from the current population. At the stage of selec-
          tion, it is necessary that a certain proportion from the entire population be
          selected, which remains at this stage of the population. The probability of
          the survival of an individual must depend on the value of its fitness func-
          tion, the selection is carried out in accordance with the criterion (30). The
          percentage of selected individuals χ is usually a parameter of the genetic
          algorithm and it is given in advance. The selection leaves $\chi \cdot (A + U + D)$
          individual that enter the next population (as a rule, these are the most
          adapted individuals), and the rest of the individuals "die".

Step 4.   Crossover and chromosome mutation. The idea behind a crossover is
          that a created "child" must share the gene information of both parents.
          $((1 - \chi) \cdot (A + U + D))/2$ pairs of individuals are selected and undergo
          the crossover and mutation according to the current crossover and obtain
          two "children" from every parent pair in the target algorithm. As a result, a
          new population consists of the same number of individuals as the previous
          one.

Step 5.   Calculation of the fitness function of a new population. If the conditions
          for the algorithm termination are not met, the fitness functions of separate
          individuals are calculated and step 3 is carried out again.

It must be noted that in the developed algorithm, the termination conditions are
set interactively. It can be either the number of iterations or a numerical parameter
that sets the value of the proximity of values in the fitness functions of the last few
populations.

The enlarged scheme of the sequence of actions while using the method of
optimizing the ShS information structure using the genetic algorithm is given in
Fig. 4.

To analyse the efficiency of the proposed method, the time of finding a solution was
compared for different variants of the fixed part of the tuple $\mathfrak{I}$, which determines the
mathematical model of the information structure, performed by the proposed method
and one of the methods for finding the optimal space on Boolean structures.

Proceeding from recommendations [17], the method of Deep Search was selected.

To implement it, matrices G, H, S were given as appropriate as corresponding
bipartite graphs. While mapping these sets on a set of information nodes, a directed
multigraph was obtained, the individual parts of which correspond to probable
options for assigning applications and data store chunks to the nodes, and the arcs
reflect the relationships between them. Acceptable options that correspond to the
above restriction are the subgraphs of this directed multigraph and the corresponding
model of such a variant can be represented by a directed graph.

The dependence of the execution time on the problem dimension was analysed,
with is equal to the solution of $\mathfrak{I}$:

$$\aleph(\mathfrak{I}) = \mathrm{card}(G) + \mathrm{card}(H) + \mathrm{card}(S) = N \cdot (A + U + D).$$

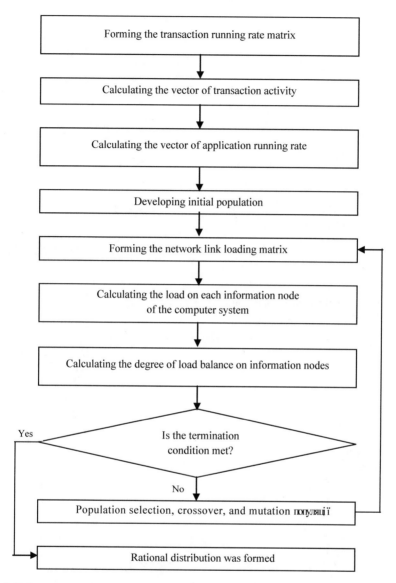

Fig. 4 Enlarged scheme of the sequence of actions while using the method of optimizing the ShS information structure using the genetic algorithm

The results of the experiment for different values of the problem dimension are generalized and shown in Figs. 5, 6 and 7 (time is presented in nominal unit).

The analysis of the experimental results showed that the proposed method gives an advantage in time over the exact methods even at $\aleph(\Im) > 60$, which can be seen in Fig. 5. Insignificant loss in time when using the proposed method at $\aleph(\Im) < 60$ is

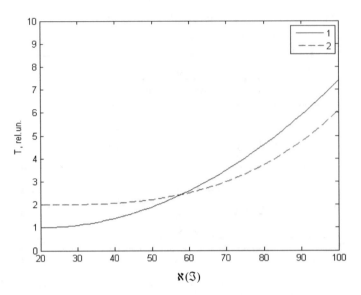

Fig. 5 Search for the optimal solution in small ShS ($\aleph(\Im) < 100$): *1*—proposed method; *2*—Deep Search method

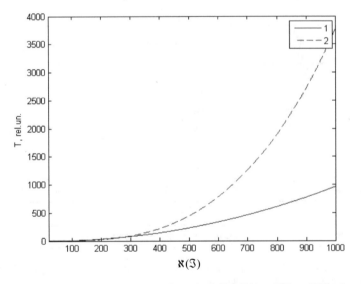

Fig. 6 Search for the optimal solution in medium-sized ShS ($100 < \aleph(\Im) < 1000$): *1*—proposed method; *2*—Deep Search method

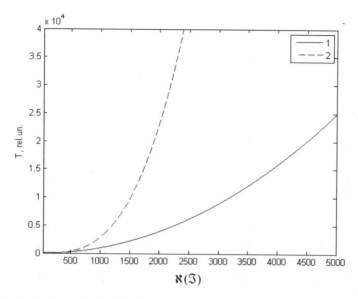

Fig. 7 Search for the optimal solution in large-sized ShS ($\aleph(\Im) > 1000$): *1*—proposed method; *2*—Deep Search method

related to the time losses of the genetic algorithm at initial stages. However, when the number of nodes increases, the advantage of the decomposition method substantially increases (Figs. 6 and 7).

The proposed method is certainly worse than the exact ones since the obtained solution is probably not optimal (but close to it). Thus, for small ShS, the results were almost the same, for medium ShS, rational solutions differed from optimal ones averagely by 5–6%, and for large ShS the loss reached almost up to10%. But, taking into account approximate assessments of the information nodes load and the exchange rate between them, such a loss can be considered insignificant as compared to the advantage in time required to find a solution.

The results obtained in previous sections enable building a model of technical structure of core network of ShS, which is isomorphic to the ShS information structure.

3 Modelling the Technical Structure of Core Network of Self-healing Systems

The technical structure of the core network of ShS focuses on active network equipment used to connect network nodes, these are intelligent and ordinary routers, switches, network adapters.

If there is a developed information structure, it is necessary to form an appropriate technical structure, that is, propose a technical node for each information node that ensures its functioning.

Based on this, it can be assumed that the number of information and technical nodes is the same.

To build the technical structure of the core network of ShS, a stratified three-layer structure is proposed. The first layer houses the control centre, the second layer contains virtual local area networks (VLANs), some of them can independently provide the operation of information nodes, others have physical components to support the operation of information nodes, which make up the 3rd layer.

Let the control centre connect to K_2 VLAN of the second layer ($K_2 \leq N$) using K_2^* network device (ND), and $K_2^* \geq K_2$, because to connect some technical nodes several network devices are needed. And K_1^* NDs provide connection with those VLANs that are isomorphic to K_1 information nodes, consequently $K_1^* \geq K_1$. To connect VLAN of the second layer with $N - K_1$ third layer component, K_3^* of the network devices is used, and $K_3^* > N - K_1$.

The schematic structure of connections between technical nodes of Core network of ShS is presented in Fig. 8.

Let us form matrices describing network connections, these matrices are Boolean, and their elements have a non-zero value when there is a physical connection between the elements:

- the matrix of connections of technical components of the second and third layers that support information nodes, with all network devices that connect such VLANs with the control centre and individual components with control VLANs:

$$Y_1^* = \left\| y_{1ij}^* \right\|, \ i = \overline{1, N}, \ j = \overline{1, K_1^* + K_3^*}; \tag{41}$$

- the matrix of correspondence of network devices that connect technical components of the first and second layers and the second and third layers:

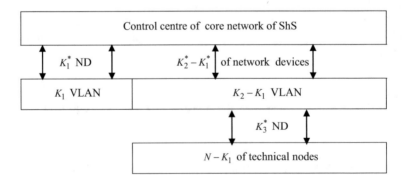

Fig. 8 Connections between technical nodes of core network of ShS

$$Y_2^* = \left\| y_{2ij}^* \right\|, \quad i = \overline{K_1^* + 1, K_3^*}, \quad j = \overline{K_1^* + 1, K_2^*} \tag{42}$$

For the elements of the matrix (41), there is a condition that the considered technical components must be connected to the upper layer by at least one network device:

$$\forall i \in \overline{1, N} \quad \sum_{j=1}^{K_1^* + K_3^*} y_{1ij} \geq 1. \tag{43}$$

Also, each technical component of the third layer, which supports the corresponding information node, must also connect to the control VLAN with at least one network device, that is:

$$\forall i \in \overline{1, K_2^* - K_1^*} \quad \sum_{j=1}^{K_3^*} y_{2ij} \geq 1. \tag{44}$$

When calculating the parameters of information streams that are transmitted through communication channels and enter network devices, the capacity of the corresponding channels must be known. Since for each layer of the technical structure there are connection matrices (41) and (42) with constraints (43) and (44), they can be used to set the bandwidth of the communication channels that provide connection.

So, let also us consider the matrices of the bandwidth of physical connections, which are determined as follows:

$$\begin{aligned} A_1^*(Y_1^*) &= \left\| a_{1ij}^* \left(y_{1ij}^* \right) \right\|, \quad i = \overline{1, N}, \ j = \overline{1, K_1^* + K_3^*}; \\ A_2^*(Y_2^*) &= \left\| a_{2ij}^* \left(y_{2ij}^* \right) \right\|, \quad i = \overline{1, K_2^* - K_1^*}, \ j = \overline{1, K_3^*}. \end{aligned} \tag{45}$$

Thus, the network technical structure is determined by the following tuple:

$$\mathfrak{I}^* = \left\langle K_1^*, K_2^*, K_3^*, Y_u^*, A_u^*(Y_u^*) \right\rangle, \quad u = \overline{1, 2}. \tag{46}$$

Isomorphism is established between information and technical nodes, but only a homomorphic relationship can be established between technical nodes and network devices. To establish isomorphism, let us combine all network devices connecting the same elements of the technical structure.

As a result, to connect the technical components that are considered with the upper layer, each technical node needs exactly one ND; while connecting the technical components of the second and third layers, exactly one combined network device is used for each element of the third layer.

And in this case

$$K_1^* = K_1; \quad K_2^* = K_2; \quad K_3^* = N - K_1$$

and respectively inequalities (43) and (44) are replaced by the following equalities:

$$\forall i \in \overline{1, N} \quad \sum_{j=1}^{N} y_{1ij} = 1; \tag{47}$$

$$\forall i \in \overline{1, N - K_1} \quad \sum_{j=1}^{N-K_1} y_{2ij} = 1, \tag{48}$$

and a rectangular matrix (41) is square, which means that isomorphism can be established between technical nodes and network devices.

This enables replacing matrix (45) with the vectors of bandwidth of physical connections:

$$\begin{aligned} A_1(Y_1) &= \|a_{1i}\|, \quad i = \overline{1, N}; \\ A_2(Y_2) &= \|a_{2i}\|, \quad i = \overline{1, N - K_1}. \end{aligned} \tag{49}$$

Within this technical structure, let us form the load matrix of joint network devices (JND) of network nodes, taking into account the ShS (according to the scheme in Fig. 5).

For JNDs that are used for any k-th information node on the third layer, the number of control VLAN is

$$\xi_k = \sum_{\zeta=K_1+1}^{K_2} (\zeta \cdot y_{2k\zeta}).$$

Then the load matrix is the following:

$$\Lambda = (\lambda_{i,j}); \quad \lambda_{i,j} = \begin{cases} 0, & if \ i = j; \\ a_{1i} + a_{1j}, & if \ i \le K_1, j \le K_1, i \ne j; \\ a_{1i} + a_{1\xi_j} + a_{2j}, & if \ i \le K_1, j > K_1, i \ne j; \\ a_{2i} + a_{1\xi_i} + a_{1j}, & if \ i > K_1, j \le K_1, i \ne j; \\ c_{2i} + c_{1\xi_i} + c_{1\xi_j} + c_{2j}, & if \ i > K_1, j > K_1, i \ne j, \end{cases} \tag{50}$$

where $\xi_k = \sum_{\zeta=K_1+1}^{K_2} (\zeta \cdot y_{2k\zeta}), i = \overline{1, N}, \ j = \overline{1, N}.$

It must be noted that in the case when the user transaction and the corresponding application belong to the same information node, there is still some load on the network equipment and communication channels in the ShS core network as requests to run the application and information on its termination are fed to the control centre.

But this service information practically does not affect the workload of both communication channels and network equipment, therefore the main diagonal in the matrix Λ is zero.

In the method for optimizing the ShS information structure given in the previous section, a matrix of exchange rate between information nodes was formed (expression (33)). The JND load can be calculated, equating square matrices Λ (33) and C (50):

$$c_{n,m} = \lambda_{n,m}, \quad m = \overline{1, N}, \ n = \overline{1, N}, \tag{51}$$

$$\lambda_n = \sum_{m=1, \, m \neq n}^{N} c_{n,m}, \quad n = \overline{1, N}, \tag{52}$$

where λ_n is the load of the n-th JND.

The number of communication channels providing communication between technical nodes of different layers is further determined, for the JND numbered n, let it be k_n. Taking into account the fact that the main criterion to distribute the elements of a computer system is the load balance, it is evenly distributed among the components and channels of one JND, that is the load of the k-th channel of JND numbered n is calculated as follows:

$$\lambda(n, k) = \frac{\lambda_n}{k_n} = \sum_{m=1, m \neq n}^{N} c_{n,m}/k_n = \left(\sum_{m=1}^{N} \frac{c_{n,m}}{k_n} \right) - c_{n,n}, \quad n = \overline{1, N}. \tag{53}$$

Taking into account the above, an extended model of the technical structure of the network can be given by the following tuple:

$$\Im = \langle K_1, N, N - K_1, Y_u, A_u(Y_u), \lambda(n, k) \rangle, \quad u = \overline{1, 2}. \tag{54}$$

So, in this section, a mathematical model of the ShS information structure is proposed as well as a method that enables forming a variant of the structure close to the optimal one in terms of the load balance criterion and proposing a model of the technical structure of the ShS core network that is isomorphic to the information structure.

4 Peculiarities of Information Transmission in Wireless Components of Self-healing Systems

4.1 Aspects of the Operation of Core Network Protocols ShS in Wireless Components

Factors such as the increase in the number of Internet resources, the trend towards an increase in the degree of subscriber mobility, the convergence of technologies to transmit various types of traffic, and mobile applications that have appeared, cause considerable interest in the efficient transmission of heterogeneous traffic using wireless networks in ShS [18, 19]. The issue of increasing the bandwidth and efficiency of such networks is becoming increasingly important.

Currently, active research is being carried out in the field of methods for qualitative and quantitative improvement of traffic transmission in wireless communication networks, which can be divided into two groups—transport and channel layers.

The TCP is the most widely used transport protocol in modern computer networks and provides guaranteed delivery of information between nodes. In addition, the functions of the TCP include segmenting and collecting data for a user, as well as stream control and congestion prevention [19, 20]. In the case of wireless networks, channel errors are predominant. By its algorithm, TCP assumes that any packet loss is due to congestion, which is true for infrastructure with reliable channels. However, in in a mobile environment, losses are mainly due to climatic conditions, physical obstacles, electromagnetic interference, the degree of mobility of end devices, and signal attenuation. To increase the protocol bandwidth in these networks, various algorithms and methods were proposed; they are implemented in such modifications as TCP Tahoe, TCP Reno, TCP SACK, NewReno, Snoop-TCP, M-TCP i I-TCP [19].

The bandwidth of wireless networks is 4–6 times lower than the bandwidth of traditional networks; at the same time, the process of transmitting information in wireless networks is significantly affected by an increased level of channel errors caused by both node motions and the physical state of the transmission medium [21]. Also, a significant number of modern studies deal with the issues of increasing the bandwidth of various wireless networks [19].

The drawbacks of the existing modifications of the TCP happen when the protocol incompletely interacts with the available network infrastructure, traffic encryption by intermediate nodes affects the network integrity, the delivery routes of data packets and acknowledgment packets do not coincide, and there is insufficient bandwidth on the route.

Currently, activities to increase the bandwidth of a transport protocol connection in a wireless network can be divided into the following categories:

- those related to breaking down connections into components along the route [19];
- those related to the influence of information encryption that is performed at the channel level [21];
- those related to modifying the TCP.

Table 1 Stack layers hierarchy of the OSI protocol and factors corresponding to them, which affect the network bandwidth

Layers	Factors
Application level	• Type of applications;
Transport layer	• Information transmission protocol; • Traffic fractality;
Network layer	• Errors caused by congestion; • Routing protocols; • Encryption methods; • Traffic fractality;
Data transfer layer	• Method of the transmission medium access; • Transmission errors probability; • Radio channel asymmetry;
Physical layer	• Obstacles; • Attenuation, diffraction, signal reflection; • Node mobility; • Errors redirection (*handover* event)

This section deals with the results of the study of approaches to the TCP modification, which is carried out exclusively at the end nodes of the connection, without affecting the intermediate elements of the network. The purpose of the section is to analyse the factors affecting the bandwidth of various implementations of the TCP when a large number of packets is lost, which is typical for wireless networks and is a key factor for ShS.

Consider the most significant factors affecting the efficiency of the TCP / IP protocol stack, and, hence, the bandwidth of the joint IP network. To analyse the TCP from the perspective of a system approach, which attempts to take into account other levels of the stack, these factors are arranged according to the hybrid model of network protocols [22], which is shown in Table 1.

The application layer of the TCP / IP stack, based on the QoS requirements of the corresponding application, determines the bandwidth required for that application.

The transport layer determines the transmission protocol, which has its own characteristics, preferred application fields, and set parameters. Selecting the implementation of the protocol and its configuration enables affecting the bandwidth [23]. The TCP transport layer provides reliable delivery of information between nodes and includes mechanisms to establish the connection, recover errors, and control the stream depending on the network load [19]. However, in wireless networks, the bandwidth of the TCP decreases significantly due to the features of such networks (the dominance of channel errors, errors during movement, and so on).

Network layer. One of the main tasks of organizing the process of transmitting information in mobile networks is to manage the mobility procedure, which provides updating routing information when mobile nodes move randomly.

Mobility management comprises two components:

- location management it is a two-step process: registering a location (updating information about the location of a node) and updating routing information that allows the network to determine the current location of a node;
- redirection management allows the network to maintain the connection established with the node during its motions and changes of access points to the network. The switching process consists of three stages (initializing, establishing a new connection, managing information streams taking into account the required level of quality of service).

For the efficient operation of the communication channel, the process of switching over a mobile node from one area to another must include [24, 25]:

- fast redirection: redirecting procedures must be fast enough to provide receiving IP packets from the mobile node from its new location without significant time delays and thereby reduce the packet transmission time as much as possible;
- continuous redirection: the redirecting algorithm must reduce packet loss to a minimum;
- efficient routing: routes between nodes must be optimized to avoid redundant transfers or detour routes;
- maintenance of a given level of service quality: the mobility management scheme must maintain quality of service standards in order to transfer various traffic properly.

To be able to establish a connection quickly to initiate information transmission when a mobile node moves, the system must have complete information about its location. If there is no such information, a node is searched for within the entire service area, which precedes the process of establishing the connection. A node that changes its network address when moving between service areas and the resulting redirection can result in losing data, increasing transmission time, and reducing bandwidth.

One of the main parameters of any communication channel is the probability of errors. There is a high probability of errors in wireless networks. If in wired networks any data loss is considered as congestion in the network, in a wireless medium, a high bit error rate is another cause for the loss of transmitted information, but the data recovery mechanisms are the same as in wired networks [26]. The inability to determine error type when transmitting information in wireless networks, and, as a result, the use of incorrect methods of dealing with congestion, reduces the network bandwidth; and a high level of errors reduces the bandwidth of the channels and increases the amount of data in the network due to repeated data transmissions. The unreliability of the information transmission channel and mobile node motion increase the number of lost data, which results in a decrease in network bandwidth.

Information transmission level. At this level, the following factors have an impact.

Media access method. One of the main tasks that is solved when building wireless networks is the task of user access rights differentiation to a constrained resource of the transmission medium. There are several basic access methods that are based on spatial, time, frequency, and code division between nodes.

Errors that occur while transmitting information cause a decrease in bandwidth; they occur when the power decreases and the signal attenuates, as a rule, when a subscriber moves away from the access point. For example, in technologies IEEE 802.11 the maximum transmission rate is maintained only in the immediate vicinity of the access point (within a radius of up to 10 m), and at a considerable distance (100 m) it is reduced to the smallest possible values [19].

The number of nodes and their location. The number of access points and the nature of their physical location in the service area affect the network bandwidth. If an access point is a common way of communication and is used as a wireless concentrator unit, the channel bandwidth assigned to each user decreases as the number of users of an individual access point increases [19].

Encryption techniques increase the information redundancy, which leads to an increase in the time needed to process data packets, which in turn results in a decrease in network bandwidth.

Radio channel asymmetry. In wireless networks, data are transmitted in two directions—from the base station to the mobile node and vice versa. These two directions are asymmetric, since when information is transmitted from the base station, the number of bit errors is less, and, therefore, the bandwidth is higher than in the opposite direction.

Physical layer. At this level, the following factors have a significant impact on the channel capacity.

Signal attenuation and interference. Signals attenuate when overcoming natural obstacles, and this leads to signal power decrease and interference, which, along with obstacles that may arise when neighbouring access points operate on the same or overlapping channels, can result in data packet loss and, consequently, bandwidth decrease.

Node mobility: when a node moves, the signal can change significantly, which negatively affects the bandwidth of the connection.

In wireless networks, the main causes of errors and bandwidth reduction are [19, 20, 23]:

- congestion;
- redirecting error (*handover* event);
- transmission channel error;
- traffic fractality.

Traffic fractality affects the operation of the network layer and information transmission layer, which leads to periodic congestion of switches and routers. The consequences of fractality can be partially compensated for at the transport level [27].

Among a number of factors causing errors and bandwidth decrease in wireless networks, one of the most significant is *handover* event, when a mobile node service is transferred from one base station to another one with the node motion. When a mobile node leaves its zone, its IP address becomes invalid. In this case, the mobile node is outside the service area, the process of exchanging information with the source is interrupted, and all the packets that are sent to the internal agent of this mobile node are deleted. To solve this problem to some extent, *mobile-IP* mechanism is used [28], which enables assigning a new address to a mobile node and registering it within the internal agent, when it enters a different service area. At the same time, all the packets that have been sent to register a new address are lost. *Mobile-IP* provides increasing stability of communication sessions. It uses two agents to transfer packets to the mobile node: the home agent (HA) and the foreign agent (FA). FA serves to transfer a new IP address of the node to the HA when a node is re-registered in another service area. As long as the node is within one service area, IP addresses do not change when transmitting information, however, after the node moves to another area, the IP address of the mobile node changes and all the packets assigned to it must have new IP addresses.

The loss of packets in the process of redirecting (*handover* event) results in:

- running congestion control algorithms (it is incorrect in this situation), which leads to a decrease in the number of transferred packets and an increase in the packet retry rate (RTO);
- long transmission delays as several timeouts necessarily occur, which increases the value of RTO. After the handover event, the sender must wait until the RTO ends to retransmit the lost packet once more. This delay period after the node has been moved causes TCP degradation;
- slow data recovery as after RTO has timed out during *handover* event, the sliding window decreases to its minimum value and only then gradually increases.

This bandwidth reduction is due to an excessive decrease in the size of the sliding window and a long wait for packet delivery acknowledgment. If, while moving, the mobile node leaves the service area, all the packets sent to it are lost. Packet losses and delays during node movements significantly affect the TCP bandwidth and, as a result, the overall network bandwidth.

A great number of methods to improve the characteristics the user traffic transmission has been recently developed and studied, which are implemented in M-TCP, WTCP, I-TCP, EBSN, ILC-TCP [19]. All of them can be divided into two large groups of solutions for the transport and channel layers. The decrease in the TCP efficiency in wireless networks is mainly due to the fact that the protocol erroneously assumes that all losses are due to network congestion. Most of the proposed ways to improve the TCP efficiency are aimed at introducing modifications that allow the TCP to distinguish between congestions and redirects. The main drawbacks of existing methods to increase the TCP efficiency in wireless networks are:

- constrained interaction with the existing network infrastructure as in the infrastructure network, intermediate network elements do not always belong to the same organizations, that is, there is not always access to such elements to make changes; all methods that break down the connection route into parts require that intermediate nodes be modified;
- traffic encryption problems since network security requires special attention, it is necessary to apply information encryption methods, for example, IPSec protocol

is an integral part of IPv6 protocol and in this case, the data are encrypted, there-fore, intermediate network nodes (base stations, routers) cannot recognize the information the packets contain, this is true for such implementations of TCP as I-TCP, MTCP and MTCP;

- divergence between information delivery routes and their acknowledgments since in certain situations, information delivery routes and their acknowledgments may not coincide; this, for example, often happens in satellite networks; in these cases, methods that break down the connection route into parts may be ineffective;
- critical sections on the route because if assume that the above points are not significant enough, there is a probability that intermediate elements become a critical section in the network.

Base stations that use protocols Snoop, I-TCP, MTCP, MTCP buffer some of the information (for example, to perform local retransmission) and partially perform additional processing of each connection passing through them. When a mobile node moves between the service areas of base stations, the information transmission process (including any data that is already buffered on the set route) must be transferred to the new base station.

On connections, broken down by TCP, when redirecting is complete, the congestion control mechanism tries to determine the bandwidth of a new channel once more. Since retransmission between the endpoints of a transport connection results in additional delay, one connection can be broken down into several serial connections. In this case, routers located at the junction of wired and wireless network segments must be used as breakpoints. The connection sharing circuit breaks the semantics of the TCP. To preserve the original semantics, an additional delay for confirmations must be introduced, which, in turn, reduces the connection speed.

Existing channel layer solutions are focused on hiding losses that occur at lower levels from the TCP. The method of local error correction at the layer of IP (Snoop TCP) is rather a solution of the channel-network layer than only channel layer. Snoop-agent is set up at the break point of the TCP connection and monitors the data and acknowledgments going through it. It also buffers unacknowledged packets, detects loss of packets by analysing re-transmitted acknowledgments and the state of local timers. Based on the information received, the agent re-transmit the lost data. Implementing this approach hides re-transmitted acknowledgments, which initiates data loss at the wireless segment of the TCP connection, preventing unnecessary attempts to recover data at the transport layer. The Snoop-agent uses the information available in TCP packets to avoid introducing additional redundancy into the frame headers of the channel layer. This solution is a sharing circuit of the output TCP connection but without breaking its semantics and also enables avoiding the conflict of re-transmissions at the channel and transport layers by suppressing retransmitted acknowledgements.

So, one of the main causes for reducing the bandwidth of a wireless network is redirecting errors that occur when a mobile node in the network service areas is re-registered, there is a significant number of channel level errors and these errors are not processed well by existing implementations of transport-level protocols.

Thus, in this section, the most significant factors affecting the bandwidth of wireless networks have been specified and analysed. Their impact on the bandwidth can be reduced by implementing mechanisms that enable recognizing the type of error (*handover* event, congestion, transmission or channel error), which makes it possible to use a more efficient mechanism corresponding to the given error type. It was specified that the bandwidth of wireless networks heavily depends on the methods and algorithms for data transmission, implemented by the transport layer protocols. The efficiency of the transport layer of the TCP / IP protocol stack for such networks can be improved by using information from other layers of the protocol stack, in particular, the service information contained in the packet that the data transmission layer and the physical layer provide, as well as by taking into account the fractality of traffic.

The results of studying the peculiarities of information transmission in ShS wireless components are given further, as well as the main expressions to calculate the efficiency of connections.

4.2   Factors Affecting the Transmission of Information in ShS Wireless Components

The efficiency of information transmission in ShS wireless components is significantly affected by delays and channel layer errors. In conditions of constant movement or re-registration of subscribers, the nature of errors is constantly changing and poorly predictable.

When analysing factors affecting the performance of wireless components, the following ones were singled out:

- a joint operation of subscribers within the same frequency range, which leads to allocating the available bandwidth among all subscribers;
- the dependence of the transmission rate on the signal power level, which causes a decrease in bandwidth when the subscriber moves away from the access point;
- incorrect determination of the needed number of access points and their irrational distribution to ensure the necessary conditions for the service area coverage;
- attenuation of the electromagnetic signal when overcoming natural obstacles;
- radio interference;
- service information and interframe interval;
- the use of additional encryption methods that increase information redundancy and time required for processing packets.

The causes for packet loss in wireless components can be the following:

- channel errors due to the deterioration of communication parameters;
- the effect of random interference that is structured and sometimes periodic;
- a break in the communication line when a mobile subscriber moves from one cell to another.

Based on the analysis, a simulation scenario was developed that illustrates the impact of the changes made on the operation of TCP.

According to Kuchuk [18], the efficiency of long-lived connection in a wireless component of ShS core network can be obtained by modelling behaviour of the mechanism to avoid congestion of a particular implementation of TCP, if assume that packet losses occur with constant rate. It is also assumed that the only cause of packet loss is network congestion. The number of packets can be obtained which were transmitted within the interval between receiving messages on packet loss. Let the average number of such packets and the average duration of such an interval be determined as Y and A respectively. Then, the average efficiency of a TCP connection can be calculated as follows:

$$B = \frac{Y}{A}. \tag{55}$$

Let two types of packet losses be considered: losses due to wireless connection and losses due to congestion.

Assume, that losses due to wireless connection occur with constant rate p, which depends on the rate of losses in a wireless channel and the packet size. The loss of packet due to congestion happens when the window size of a TCP reaches W_M value that can be calculated as:

$$W_M = \frac{C_d}{S} \cdot RTT_{\min} + M, \tag{56}$$

where $RTT_{\min}$ is minimum time to transfer a packet within an established connection; C_d is the size of the wireless connection bandwidth; M is a buffer size.

Thus, the peculiarities of information transmission in wireless components of core network of ShS are studied. It is shown that delays and errors of a channel level affect significantly the efficiency of information transmission. Under the conditions of constant motion of subscribers, the error character changes continuously and can be predicted poorly. Also, the most influential factors of packet loss are specified and the main expressions to calculate the efficiency of connections are given.

5   Modelling the Traffic of Wireless Components of Core Network of Self-healing Systems

The mostly widely used model of telecommunication traffic that has fractal features of is an ON/OFF-model that is also widely used to explain physical causes of fractal phenomena in current telecommunication networks [29].

The traditional ON/OFF-model forms a process that is a process of incrementing fractal Brownian motion. In its nature, it is a process that alternates between two states—0 or 1. Time T_0 (the time in "0" state) is a random variable with a probability density function $w_0(t)$, and time T_1 (the time in "1" state) is a random variable with a probability density function $w_1(t)$; $w_0(t)$, $w_1(t)$ are characterized by the distribution

$$w_i(t) \sim t^{-(\alpha_i+1)}, \qquad (57)$$

where $i = 0, 1$; $\alpha_i \in (1, 2)$.

Based on the analysis of obtained time series $\{w_i(t), \ t \geq 0\}$, ON/OFF-models of individual sources traffic are built, which illustrate the periods of this process activity. However, it is not taken into account that a packet is transferred in groups in the periods of activity of each individual traffic source [30]. The peculiarities of wireless components of core network of Self-healing Systems are not also taken into account.

So, it is necessary to develop an ON/OFF-model of traffic focused on the peculiarities of core network of Self-healing Systems. First, consider the traffic model for an individual source.

5.1 Developing an ON/OFF Model of Individual Source Traffic of a Wireless Component of ShS Core Network

The first of the above factors is considered in [18] by broadening a traditional ON/OFF-model by the structure for accounting the periods of source activity. To achieve this, for random j-th source $\left(j = \overline{1, N}\right)$, the interval of time $[0, T]$ is partitioned into $n_j^{(on)}$ periods of the source activity $i = \overline{1, n_j^{(on)}}$ is given (Fig. 9):

$$R_j^{(on)} = \left\{ t_{j,i}^{(on)} = \left\langle t_{j,i}^{(0)}, \ \Delta t_{i,j} \right\rangle \middle| i = \overline{1, n_j^{(on)}}; \right.$$
$$\left. t_{j,i+1}^{(0)} > t_{j,i}^{(0)} + \Delta t_{j,i}; \ t_{j,i}^{(0)} \in \left[0, T; \ t_{j,n_j^{(on)}}^{(0)} \leq T \right] \right\}, \qquad (58)$$

where j is a source index; i is the number of activity of the j-th source (ON-period); $t_{j,i}^{(0)}$ is the beginning of the i-th ON-interval with the duration of $\Delta t_{j,i}$.

This partitioning enables taking into account the order in which groups of packets pass each ON-interval of a given source by introducing the following refinement partitioning

$$R_{j,i}^{(A)} = \left\{ t_{j,k}^{(A)} = \left\langle t_{j,i,k}^{(0)}, \ \delta t_{j,i,k} \right\rangle \middle| k = \overline{1, n_{j,i}^{(A)}}; \ t_{j,i,k}^{(0)} \in R_{j,k}^{(on)} \right\}, \qquad (59)$$

where k is the number of the interval for which the group of packets is transmitted into the activity interval $t_i^{(on)}$; $n_{j,i}^{(A)}$ is the amount of groups of packets with the initial time

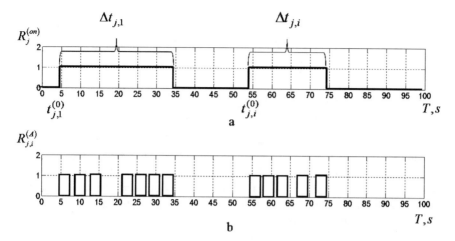

Fig. 9 Hierarchical structure of an ON/OFF-model of traffic of the critical section of the wireless component of ShS core network: **a** the activity intervals of the *j*-th source; **b** the intervals of packet group transfer in *i*-th activity intervals

$t_{j,i,k}^{(0)}$ and the duration of transmission $\delta t_{j,i,k}$, which are transferred into the activity interval $t_i^{(on)}$ by the j-th source.

Such basic partitioning of the time axis for the ON/OFF model takes into account the structure of the source activity periods.

Thus, the model takes into account two of the above reasons that determine the properties of traffic fractality: user behaviour and traffic generation. In addition, the model enables simple coordination with the protocol of system auto recovery.

However, the above expressions have a single amplitude and characterize the behaviour of only one user, but do not take into account handover, which is typical for a wireless network.

5.2 Accounting Handover in an ON/OFF Model of Individual Source Traffic

The node mobility in a wireless network may cause a handover [19]. Since the node location is directly related to its IP address, therefore, the fact that the IP address changes when the mobile node changes the base station (network access point) requires notification. Handover is an event that can occur when a mobile node leaves the service area of one access point due to its motion into the service area of another point, and leads to a temporary termination of traffic generation by the transport layer of a protocol stack, while maintaining all transmission parameters, and the subsequent restoration of such generation according to saved parameters.

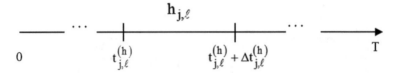

Fig. 10 The ℓ-th interval of a hadover models

Correct processing of such an event allows the mobile unit to maintain connection continuity while it is moving and changing access points to the network.

To allow an ON/OFF model of individual source traffic to take into account a handover at the input of a wireless component of Core network of Self-healing Systems, assume that within the time interval $[0, T]$ the j-th node left the service area of a wireless component of core network L times, and the ℓ-th interval of a handover $(\ell = \overline{1, L})$ is given as $h_{j,\ell} = \left(t_{j,\ell}^{(h)}, t_{j,\ell}^{(h)} + \Delta t_{j,\ell}^{(h)} \right)$ (Fig. 10).

This enables determining the function of a time shift in an ON/OFF model that allow the handover effect to be considered:

$$h_j(t) = t + \sum_{t_{j,\ell}^{(h)} < t} \Delta t_{j,\ell}^{(h)}, \ t \in [0, T], \ \ell \in [1, L]. \tag{60}$$

Taking into account function (60), partitioning (58) and (59) enable taking into account the peculiarities of a wireless component of ShS core network ShS as follows:

$$R_j^{(BP,on)} = \left\{ h_j\left(t_{j,i}^{(on)} \right) = \left\langle h_j\left(t_{j,i}^{(0)} \right), \Delta t_{i,j} \right\rangle \middle| i = \overline{1, n_j^{(on)}}; \right.$$
$$h_j\left(t_{j,i}^{(0)} \right) \in \left[0, T; h_j\left(t_{j,n_j^{(on)}}^{(0)} \right) \le T \right] \right\}; \tag{61}$$
$$\left. h_j\left(t_{j,i+1}^{(0)} \right) > h_j\left(t_{j,i}^{(0)} \right) + \Delta t_{j,i};$$

$$R_{j,i}^{(BP,A)}$$
$$= \left\{ h_j\left(t_{j,k}^{(A)} \right) = \left\langle h_j\left(t_{j,i,k}^{(0)} \right), \delta t_{j,i,k} \right\rangle \middle| k = \overline{1, n_{j,i}^{(A)}}; \ h_j\left(t_{j,i,k}^{(0)} \right) \in R_{j,k}^{(BP,on)} \right\}. \tag{62}$$

Partitioning (61) and (62) allow the handover effect in the wireless component of the core network to be considered automatically; this effect is formalized by function (60) while building an ON / OFF model.

It should be noted that when there is no handover (L = 0), there is no time shift does not, that is $h_j(t) = t$, and, consequently,

$$R_j^{(BP,on)} = R_j^{(on)}; \quad R_{j,i}^{(BP,A)} = R_{j,i}^{(A)}.$$

The development of an ON/OFF model of individual source traffic at the input of the wireless component of core network of Self-healing Systems, which takes into account the channel bandwidth allocation is given in the following section.

5.3 Channel Bandwidth Allocation

Knowing the speed of each individual source, the values of traffic rate generated by it at different time intervals can be predicted. However, this rate is not constant due to the methods of bandwidth allocation based on the feedback between the source and destination [31, 32]. To do this, while building an ON/OFF model, it is necessary to approximate the changes of traffic sources speed by a periodical function. Therefore, in fact it is possible to speak about the process periodicity (or about quasi periodicity), which is due to using the method of bandwidth allocation based on feedback.

At the transport layer of the OSI model, the method to control the speed at which traffic is sent by the source is used, which is based on the resizing of the floating window of the latter.

Figure 11 illustrates the process of additive increasing and multiplicative decreasing the floating window size at a transmission rate of 32 Mbit/s.

Since the fluctuations in the traffic rate from one source reach on average the values from 8 to 12% of the value of the bandwidth allocated for the connection, the frequency of phases sequence should bring insignificant but tangible correlations into the traffic generated by the source.

It is obvious that these speed changes are similar to fluctuations and they can be approximated by a periodical function (Fig. 12a). In fact, it is possible to speak about the process periodicity (or about quasi periodicity), which is due to using the method of bandwidth allocation based on feedback [33].

Each source can be modelled by periodical or quasi periodical signals with the values of amplitudes A_j and frequencies ω_j. When combining an ensemble of quasiperiodic sources, random coincidences close to the maximum or minimum values of amplitudes are like bursts or decays in intensity in the combined process against the background of longer, smoother sections on the time axis.

In addition to the variations of the considered values, it is necessary to take into account the variations in the beginning of the activity period of each source within the stochastic phase that is similar to the variation in the connection time of various individual sources.

Since Heaviside function [30] for the case of $\delta t < \Delta t$ have the property of

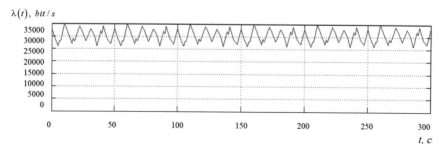

Fig. 11 The process of additive increasing and multiplicative decreasing the floating window size

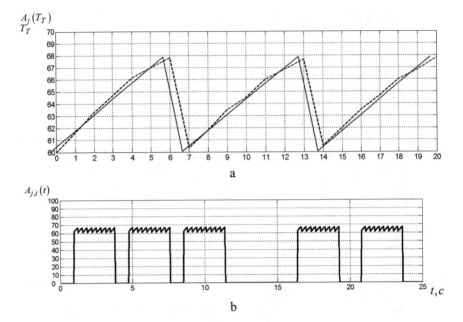

Fig. 12 Modelling an individual source traffic: **a** speed change approximation; **b** the model of individual source traffic within the activity period

$$[Hv(t) - Hv(t + \Delta t)] \cdot [Hv(t) - Hv(t + \delta t)] = Hv(t) - Hv(t + \delta t),$$

and taking into account that $T_T \ll \delta t < \Delta t$ (where T_T is the time of packet transmission within the set route), the change in transmission rate due to feedback control between destination and traffic source within the given partitioning (5), (6), can be described as follows:

$$
\begin{aligned}
C_{j,i,k}(t) = \sum_{\ell=0}^{\ell_{j,i,k}} & \left(\left(A_j \cdot (1 - \alpha_{j,i}) + (t - \ell \cdot \tau_{j,i,k}) \cdot \frac{\omega_{j,i,k} \cdot \gamma_{j,i,k}}{T_T} \right) \right. \\
& \times \left(Hv \cdot (\tau_{j,i,k} \cdot \ell + \omega_{j,i,k}) - Hv \cdot (\tau \cdot \ell + \omega_{j,i,k} + T_T) \right) \\
& + \left(\gamma_{j,i} \left(t - h_j \left(t_{j,i,k}^{(0)} \right) \right) + A_j \cdot (1 - \alpha_{j,i}) - \ell \cdot \tau_{j,i,k} \right) \\
& \times \left. \left(Hv \cdot (\ell \cdot \tau_{j,i,k}) - Hv \cdot (\ell \cdot \tau_{j,i,k} + \omega_{j,i,k}) \right) \right),
\end{aligned}
\tag{63}
$$

where $h_j \left(t_{j,i,k}^{(0)} \right)$ is the moment the process starts; $\gamma_{j,i,k}$ is the factor of the source rate change; $\ell_{j,i,k}$ is the number of intervals of the transmission rate increase of the j-th source at the i-th activity interval within $\left[h_j \left(t_{j,i,k}^{(0)} \right), h_j \left(t_{j,i,k}^{(0)} \right) + \delta t_{j,i,k} \right]$; A_j is available rate for the j-th source; $\alpha_{j,i}$ is a factor that takes into account variations of the transmission rate change;

$$\omega_{j,i,k} = \frac{2\alpha_{j,i} \cdot A_j}{\gamma_{j,i,k}};$$

$$\tau_{j,i,k} = h_j\left(t_{j,i,k}^{(0)}\right) + \omega_{j,i,k} + T_T.$$

The above expressions (61), (62) as well as (63) describe the traffic behaviour of an individual source within the particular ON-section of continuous traffic generation.

The analysis of the aggregated process from many sources with characteristic quasiperiodic components enables determining the fact of the effect of the transmission rate control on the formation of fractal properties of traffic [34]. Combining an ensemble of such sources, provided that the statistical characteristics are identical, enables predicting traffic behaviour for some time in advance. To do this, it is necessary to develop a combined model of traffic at the input of the wireless component of the core network of Self-healing Systems, taking into account all of the listed reasons of fractality when there are many sources. This model should focus on the Self-healing Systems monitoring and detecting mechanism and support the system auto-recovery protocol.

5.4 Developing an ON/OFF Model of Summary Traffic of a Wireless Component of Core Network of Self-healing Systems

The ON/OFF model of an individual j-th source at the i-th activity interval that takes into account both the hierarchical structure of its activity periods and the variations of rate due to the methods of bandwidth allocation within interval $\left[h_j\left(t_{j,i}^{(0)}\right), h_j\left(t_{j,i}^{(0)}\right) + \Delta t_{j,i}\right]$, proceeding from expression (62), is as follows:

$$I_{j,i}(t) = \sum_{k=0}^{n_{j,i}^{(A)}} C_{j,i,k}(t), \tag{64}$$

and for the integrated input stream, the rate of aggregated traffic of a wireless component of core network of Self-healing Systems is calculated as follows:

$$I(t) = \sum_{j=0}^{N} \sum_{i=0}^{n_j(on)} I_{j,i}(t) \tag{65}$$

with the following constraints:

$$I(t) \leq I_{max}; \tag{66}$$

$$h_j(t) = t + \sum_{t_{j,\ell}^{(h)} < t} \Delta t_{j,\ell}^{(h)}, \quad \ell \in [0, L], \quad t_{j,0}^{(h)} = T; \quad \Delta t_{j,0}^{(h)} = 0; \tag{67}$$

$$\sum_{i=0}^{n_j(on)} \Delta t_{j,i} \leq T_j; \tag{68}$$

$$\sum_{k=0}^{n_{j,i}^{(A)}} \delta t_{j,i,k} \leq \Delta t_{j,i}; \tag{69}$$

$$\alpha_{j,i} << 1, \tag{70}$$

$$T_T \cdot l_{j,i,k} < \delta t_{j,i,k}; \tag{71}$$

$$j \in \overline{1, N}; \quad i = \overline{1, n_j^{(on)}}; \quad k = \overline{1, n_{j,i}^{(A)}},$$

where T_j is time needed to transmit data from the j-th source through the critical section; I_{max} is the maximum allowable transmission rate at the critical section of the wireless component of ShS core network.

The obtained mathematical model (65)–(71) that describes both data streams from one source and the aggregated traffic, unlike analogues, takes into account actual processes in joint networks of data transmission including a wireless component of core network of Self-healing Systems.

5.5 Modelling the Process of Bandwidth Reallocating at the Critical Sections of a Wireless Component of Core Network of Self-healing Systems

While predicting the changes of traffic rate in a wireless component of ShS core network, the time to data transmission at the critical network section (CNS) can be reduced by using this model.

The congestion in network devices affect the time of data transmission in wireless networks, which should be considered in the context of basic requirements data transmission networks must meet [35, 36]. Since the congestion both on individual routers, as well as in a wireless component of ShS core network is a rather complex process and often cannot be formalized, there is a need for a more specific description of the process of circulation of data streams at critical sections within established routes in wireless components of ShS CN as well as time delays in packet transmission (average data packet transmission time) related to them [37]). Therefore, the most complete characteristic of the congestion process in routers of a wireless component of ShS CN can be obtained by considering it along with the average transmission time of a data packet. Considering that the bandwidth of the set route with critical sections where there is constrained bandwidth is determined by the bandwidth of such critical sections, it is necessary to consider only the transmission time of a data packet to wireless components of ShS CN.

According to studies [38–41], data streams in modern networks, including wireless ones, are characterized by a high factor of intensity peak values deviation. Thus, when combining a large number of data streams, the process is not smoothed by averaging over time intervals, that is, while resource allocating the possibility of bursts of traffic intensity is not taken into account. However, short periods of high traffic intensity are followed by long periods of low intensity. Considering that the congestion is usually controlled based on the analysis of packet loss during congestion, which reduces the transmission rate, and the burst in intensity may be short, reducing the transmission rate is not always justified. As a result, the connection bandwidth is not used completely. Taking into account the fact that traffic at the input of a CNS of a wireless component of CN ShS CN cab be fractal and, as it is shown in previous sections, can be predicted due to the correlation dependence that is a feature of fractal processes, this feature can be used to predict the bandwidth size required for transmission.

Figure 13 illustrates how the required bandwidth is statically set and how the bandwidth can be dynamically changed by predicting changes in traffic rate.

Fig. 13 Principle of bandwidth dynamic change based on predicting the traffic rate: **a** statically set bandwidth; **b** dynamically set bandwidth

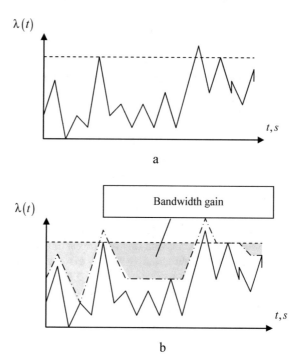

Fig. 14 Changes in the ratio between service traffic (ST) and information traffic (IT) at a critical section (CS) of a network

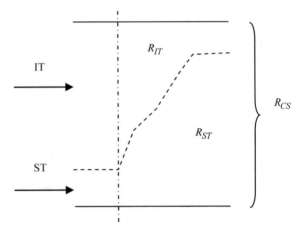

Therefore, to develop the method of reallocating bandwidth at CNS of wireless components of ShS CN, it is necessary to develop the method of predicting the changes in traffic rate and the method of bandwidth reallocation. However, it should be taken into account that when a network operates properly, the service traffic (ST) occupies 5–10% of the general bandwidth, but in the context of dynamic changes in the network topology, when there are segments that are critical sections (CS) in a wireless component of ShS CN, the service traffic might occupy 80% and even more of the general bandwidth at the given section (Fig. 14) [40].

This results in decreasing the transmission rate of information traffic packets, leads to their loss, and, consequently, to increasing the data transmission time. Hence, there is a need to reallocate the bandwidth at CNS of a wireless component of ShS CN between the streams of service traffic and information traffic (IT) to provide the transmission of information traffic packets and network operation under the conditions of changes in its structure and composition.

Consider the process of reallocating the bandwidth at a CNS of a wireless component of ShS CN. The available bandwidth for each connection is found using methods based on feedback between the source and the destination. Protocols through algorithms based on these methods determine the distribution point (DP) based on dropping packets when the stream exceeds the available bandwidth.

Consider the operation of modes of AIMD method in detail.

1. Quick start phase. The objective of this phase is to determine the maximum allowable transmission rate for the established connection as soon as possible. Data transmission rate increases exponentially. When the maximum value, which is the sum of the value of the available bandwidth and the allowable buffer memory space because receiving the packet is not confirmed, is reached, the mode of the multiplicative reduction of the rate of data transmission to the network is enabled [41].

2. Within the multiplicative reduction phase the rate of data transmission to the network is deliberately set below the available network bandwidth, and the recovery phase is enabled.
3. Within the recovery phase the rate of data transmission to the network increases linearly to the value of the bandwidth, and then the fine adjusting mode is enabled. Next, the floating window size and its multiplicative reduction are set by the sequence of the additive increase. All this time, packet loss occurs when the available bandwidth is exceeded, which leads to an increase in data transfer time.
4. Within the congestion avoidance phase when the time required for receiving acknowledgment is expired, as well as in the situation when there is no time for acknowledgment, the size of the floating window is reduced to its minimum value.

To reduce the data transmission time and set the maximum size of a floating window, when traffic has the features of fractality, predicting is possible based on the proposed ON-OFF model. Figure 15 illustrates the process of determining such a DP.

Assume that R is the bandwidth of a wireless component of ShS CN, R_{ST} is the bandwidth given to the service traffic (ST), R_{IT} is the bandwidth of CNS, given to the information traffic (IT); $\left(R_{IT}^{(0)}, R_{ST}^{(0)} \right)$ is the initial (at the moment of time t_0) allocation of CNS bandwidth under the condition $R_{IT}^{(0)} + R_{ST}^{(0)} < R$.

Existing methods find DP streams $\left(R_{IT}^{(x)} + R_{ST}^{(x)} \right)$ by sequentially changing the phases of an additive increase in the size of the floating window and its multiplicative decrease in case of packet loss. Given the ST priority, the use of such methods results in a disproportionate bandwidth allocation. Each increase in the transmission rate of

Fig. 15 Finding the distribution point for each connection based on the feedback between the source and destination

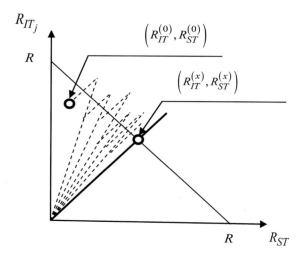

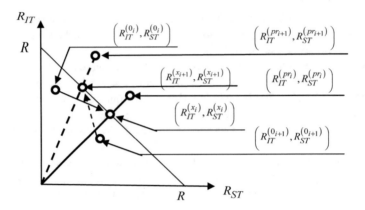

Fig. 16 Finding a DP for each connection based on predicting the information traffic rate

the bandwidth available to the connection leads to packet loss, and if there are fractal properties of incoming streams, the process of determining a DP becomes lengthy.

The proposed method does not suggest determining a DP based on predicting the intensity values of individual streams that pass through a critical section but based on the sum information traffic using the developed ON/OFF model, which enables determining maximum values of IT intensity within the interval of prediction. To do this, statistical characteristics of input are analysed and their features of fractality are checked. If the value of the Hurst exponent indicates the traffic fractality ($0.75 \leq H < 1$), the prediction is carried out using the developed advanced ON/OFFmodel of traffic at the input of a critical section of a wireless component of ShS CN. Otherwise, a DP is determined by the mean values of ST or IT, or the existing methods to find DP are used.

After predicting, the predicted ratio of IT and ST is determined as $\left(R_{IT}^{(pr_i)}, R_{ST}^{(pr_i)} \right)$.

Taking into account R_{Ky}, DP $\left(R_{IT}^{(x_i)}, R_{ST}^{(x_i)} \right)$ is determined by proportionally predicted values (Fig. 16):

$$R_{IT}^{(x_i)} = \frac{R}{R_{IT}^{(pr_i)} + R_{ST}^{(pr_i)}} \cdot R_{IT}^{(pr_i)}; \quad R_{ST}^{(x_i)} = \frac{R}{R_{IT}^{(pr_i)} + R_{ST}^{(pr_i)}} \cdot R_{ST}^{(pr_i)}. \quad (72)$$

Subsequent points for subsequent intervals ($i + 1$, etc.) are determined similarly. Stepwise changing the ratio of ST and IT is shown in Fig. 17.

A feature of the proposed method is that there is no priority in determining the DP at the critical section and that the bandwidth is distributed proportionally between the ST and IT streams. The effect of a probable increase in the delay time of ST packets can be disregarded, since the network response time to a topology change is an order more than the time a packet needs to pass through a critical section.

The algorithm of traffic rate *control at a wireless component of* ShS CN *based on changing the floating window size.*

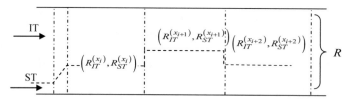

Fig. 17 Ratio of the bandwidth of the given IT and ST at a critical section of a wireless component of ShS CN when the reallocation is proportional

The developed method of proportional bandwidth allocation based on the developed ON/OFF model of traffic at the input of acritical section of a wireless component of ShS CN in which, unlike analogues, a distribution point between service and information traffic provides proportional bandwidth allocation and can be implemented by the following algorithm.

Step 1. Statistical characteristics of data streams are analysed based on the monitoring of data streams at the time interval T at the input of a wireless component of ShS CN and each of them is classified according to the criterion of the value of the Hurst exponent.

Step 2. If the Hurst exponent for both traffics is $H \leq 0.75$, ST and IT are distributed according to the known control methods that are used in the given wireless component of CN ShS CN. Further monitoring is carried out at Along with this, further monitoring is carried out at the next time interval.

Step 3. If the Hurst exponent is within the limits of $0.75 < H < 1$, probable maximum values of traffic rate are predicted based on the developed ON/OFF model (65)–(71).

Step 4. A DP is determined for the values being predicted by expression (72).

Step 5. If the sum of maximum values of ST and IT rates does not exceed the value of sum bandwidth of a wireless component of ShS CN, the CNS bandwidth is allocated proportionally according to the values being requested.

Step 6. If the sum of maximum values of ST and IT rates exceeds the value of sum bandwidth of a wireless component of ShS CN, a point of a bandwidth proportional allocation at a wireless component of ShS CN is calculated by expression (72).

Step 7. When DP is determined for those streams that cumulatively exceed the value of the proportionally distributed bandwidth, in accordance with the congestion acknowledgment procedure (value "1" is given to the current packet in the field provided by the control protocol), the sources receive the packet delivery acknowledgment along with the command to reduce the transmission rate up to the level set in the protocol. Then the transmission rate increases according to the additive increase phase.

Step 8. If further monitoring of the streams at the input of a wireless component of ShS CN indicates that the Hurst exponent for them takes value of $H \leq 0.75$, the existing methods of bandwidth allocation are enabled. Along with this, further monitoring is carried out at the next time interval.

The offered method and the algorithm of bandwidth reallocation between ST and IT traffic can be modified to redistribute IT components.

Assume that CNS serves K of traffic streams within the considered time interval. Using the offered model, the required bandwidth at the i-th time interval $R_{IT,k}^{pr_i}$ is predicted for each of K streams. Next, for each stream weight factor α_k is determined, this factor specifies the priority of its passage at the given section ($\sum_{k=1}^{K} \alpha_{k=1}$). It should be noted that nonpriority streaming is preferred at CNS, that is $\alpha_k = 1/K \quad \forall k$.

Using the above approach to bandwidth proportional allocation between IT and ST, the bandwidth is determined for each service stream

$$R_{IT,k}^{(x_i)} = \frac{\alpha_k \cdot R_{IT,k}^{(pr_i)}}{\sum_{k=1}^{K} \left(\alpha_k \cdot R_{IT,k}^{(pr_i)} \right)} \cdot R_{IT,k}^{(x_i)}.$$

The Hurst exponent is assessed in each IT stream to make predictions. If this values does not meet the condition of $0.75 \leq H < 1$, the known methods of bandwidth allocation are used, otherwise the rate values λ_{pr} for these streams are predicted by the developed ON/OFF model.

While studying Self-healing Systems, there is a wide range of tasks that are intimately connected with the life cycle of control systems for various objects. Therefore, the models of these systems evolution are of great interest and they are considered in the following section.

6   Modelling the Process of Self-healing Systems Evolution

6.1   Mathematical Model of the Evolution Process of the Topological Structure of Core Network of Self-healing Systems

Currently, a wide range of tasks that are intimately connected with the life cycle of control systems for various objects has been insufficiently studied since there are many factors, approaches, and criteria used. Such tasks are of particular importance to monitor and control critical infrastructure entities.

When operating, ShS must quickly respond to changes in the topology of computer networks, especially when some parts of the network are destroyed, when the system is urgently expanded with redundant components.

Formalizing tasks related to planning the evolution of topological structures of computer networks usually includes building models that take into account the dynamics of the development of specific components. The main problem arising while building such models is to choose the best way to formalize the process of the dynamic evolution of ShS CN components. Therefore, it is necessary to develop approaches to modelling the evolution of the topological structure of the core network of Self-healing Systems.

While modelling the process of evolution of the topological structure of ShS CN, the basic components are data sources, information addressees, as well as the number and nature of the connection between them. So, while formalizing tasks for ShS components, probable scenarios of their evolution should be set as well as the characteristics of data streams linked to them.

When formalizing the task, in the most general case, there exist two probable evolution scenarios—fixed and controlled ones. The fixed scenario assumes a priori guiding the activity of the ShS control system and is intimately connected with the technologies used for evolution. When implementing a controlled scenario, when evolution is planned dynamically and there is a number of external constraints of a different nature, it is most rational to find a solution within a certain class of specific vector functions, which, in turn, can significantly complicate the task formalization and solution and can cause step out from the technologies that are really available to implement such an evolution.

To find the optimal option of ShS CN evolution, specific options for the evolution of components should be singled out from a variety of probable options under existing constraints and certain time events, as well as the nature and dynamics of existing data streams.

At the first stage, a mathematical model for the evolution of ShS CN nodes is proposed. The evolution process is examined at time interval $[0, T]$ that is divided into M time segments of probable development of network topology:

$$[0, T] = \bigcup_{m=1}^{M} [\tau_{m-1}, \tau_m], \quad \tau_0 = 0, \quad \tau_M = T. \tag{73}$$

Consider J network autonomous components (each j-th component $(j = \overline{1, J})$ assumes the capability of autonomous input and partial processing of information). Each component includes I_j input nodes and V_j processing nodes, moreover, these sets can overlap.

Introduce the following variables:

$z_{i_j m}$ is the cost to develop an input node i_j $(i_j = \overline{1, I_j})$ within m period;

$z_{v_j m}$ is the cost to develop a processing node v_j $(v_j = \overline{1, V_j})$ within m period;

P_{jm} is a computing resource reserve of j component to process data at the beginning of m period;

$P_{i_j m}$ is a computing resource size required to process additional data obtained as a result of i_j node development within m period;

ℓ, $(\ell = \overline{1, L})$ is one of the probable options of processing nodes;

$P_{v_j m}(\ell)$ is an increase of the computing resource of v_j node within m period within the development option ℓ.

Introduce the following Boolean variables:

$x_{i_j m} = 0$, if i_j node does not develop within m period, otherwise $x_{i_j m} = 1$;

$y_{v_j m} = 0$, if v_j node does not develop within m period, otherwise $y_{v_j m} = 1$.

The general level of cost to develop j component within the considered period is

$$Z_j = \sum_{m=1}^{M} \left(\sum_{i_j=1}^{I_j} z_{i_j m} \cdot x_{i_j m} + \sum_{v_j=1}^{V_j} z_{v_j m} \cdot y_{i_j m} \right). \tag{74}$$

If the resource of means to develop a computer network does not exceed z_{set}, then

$$\sum_{j=1}^{J} Z_j \leq z_{set}. \tag{75}$$

To determine the option to develop processing units, the inequality is used, which provides the capability of processing additional load:

$$P_j^{(0)} = \sum_{v_j=1}^{V_j} P_{v_j m}(\ell) \geq H \left(\sum_{i_j=1}^{I_j} P_{i_j m} - P_{jm} \right) \forall m, \tag{76}$$

where H is Heaviside function. The reserve values are modelled recurrently:

$$P_{j(m+1)} = H \left(P_{jm} - \sum P_{v_j m}(\ell) \right). \tag{77}$$

When constraints (75)–(77) are met, the option of the optimal use of available means can be considered as a criterion. Then the objective function of the task optimization is as follows:

$$P_j^{(0)} \to \max. \tag{78}$$

At the second stage, the evolution of transit nodes of ShS CN is modelled.

The dynamics of the evolution of connections and the characteristics of data transmitted with their help have the greatest effect on the evolution of a computer network topology. The difference in rates between the output and input data streams in j component within m period can be calculated as:

$$Q_{jm} = Q_{jm}^0 + Q_{jm}^u - Q_{jm}^6, \tag{79}$$

where Q_{jm}^0 the difference value in the initial period; Q_{jm}^u is the rate of the output data stream; Q_{jm}^6 is the rate of the input data stream.

Introduce the following variables:

$x_{(jj')m}$ is the information traffic rate between components j and j' within m period;

$\hat{\theta}$ is a set of connections in a computer network in which the transmission direction is severely regulated;

$\tilde{\theta}$ is a set of connections in a computer network in which the transmission direction is not known a priori;

θ'_j is a subset for all elements of which the second index in the pair $(j'j)$ is index j;

θ''_j is a subset for all elements of which the first index in the pair (jj') is index j;

$z^\ell_{(jj')m}$ is a Boolean variable which takes on the value one only in the case when between components j and j' while planning m, the implementation of an intermediate component of ξ $(\xi = \overline{1, L_{(jj')}}, \ (jj' \in \theta))$ type started;

$s^\xi_{(jj')}$ is the bandwidth of intermediate component (jj') of ξ type.
It can be stated that during each period of planning m, it is true for all the components of a computer network that the sum amount of data that are sent is not less than the amount of data that are delivered and, consequently, the following inequality is true:

$$\sum_{j=1}^{J} Q_{jm} \leq 0. \tag{80}$$

Therefore, there is such a balance ratio for j component while planning evolution m:

$$\sum_{(jj')\in\theta'_j} x_{(jj')m} - \sum_{(jj')\in\theta''_j} x_{(jj')m} = Q_{jm}. \tag{81}$$

Constraints on information stream rate between two components within each evolution planning periods are the following:

$$\sum_{\xi=1}^{L_{(jj)}} \sum_{m=1}^{t-M^\xi_{jj'}} s^\xi_{(jj')} z^\xi_{(jj')m} \geq x_{(jj')m} + x_{(j'j)m}, \ (jj') \in \theta. \tag{82}$$

Introduce the following variables:

R_m is a sum resource allocated to intermediate components while planning m;

$R^\xi_{(jj')\eta}$ is resource cost to implements the network section of ξ type between a pair of components (jj') within period η;

$M^{\xi}_{(jj')}$ is the duration of the implementation of a transport component of ξ type between a pair of components (jj').

The condition of constrained resource consumption within planning m can be written:

$$\sum_{(jj')\in\theta}^{L_{(jj')}} \sum_{\xi=1}^{L^{\xi}_{(jj')\eta}} \sum_{\eta=1} R^{\xi}_{(jj')\eta} z^{\xi}_{(jj')m-\eta+1} \leq R_m, \quad L^{\xi}_{(jj')m} = \min\left\{M^{\xi}_{(jj')}, m\right\}. \tag{83}$$

In the context of optimizing a computer network evolution, the task assumes solving two related problems: optimal planning for the development of components and optimal planning for the development of the network. In this case, the objective function with the constraints (80)–(82) is as follows:

$$\sum_{j=1}^{J}\sum_{m=1}^{M}\left(\sum_{i_j=1}^{I} K_{i_jmj} + \sum_{v_j=1}^{V} K_{v_jmj}\right)$$
$$+ \sum_{m=1}^{M}\left(\sum_{(jj')\in\theta} Q_{(jj')m}x_{(jj')m} + \sum_{(jj')\in\theta}\sum_{\xi=1}^{L_{(jj')}} K^{\xi}_{(jj')m} z^{\xi}_{(jj')m}\right) \to \min, \tag{84}$$

where $Q_{(jj')m}$ is specific operating cost within planning m, $K^{\xi}_{(jj')m}$ is the capital cost for the implementation of an intermediate component of ξ type between components j and j'; K_{imj} is the capital cost for the implementation of i node of integrated component j; K_{vmj} is the capital cost for the implementation of v node of integrated component j (is calculated while solving tasks (81)–(84)).

However, the following constraints should be taken into account:

• balance ratio

$$\sum_{(jj')\in\theta'_j} x_{(jj')m} - \sum_{(jj')\in\theta''_j} x_{(jj')m} = Q_{jm},$$

• information stream rate

$$\sum_{\xi=1}^{L_{(jj)}} \sum_{m=1}^{t-M^{\xi}_{jj'}} s^{\xi}_{(jj')} z^{\xi}_{(jj')m} \geq x_{(jj')m} + x_{(j'j)m}, \quad (jj') \in \theta;$$

• constrained resource consumption

$$\sum_{(jj')\in\theta} \sum_{\xi=1}^{L_{(jj')}} \sum_{\eta=1}^{L^{\xi}_{(jj')\eta}} R^{\xi}_{(jj')\eta} z^{\xi}_{(jj')m-\eta+1} \leq R_m, \quad L^{\xi}_{(jj')m} = \min\left\{M^{\xi}_{(jj')}, m\right\},$$

where R_m is a sum resource allocated to intermediate components within planning m; $R^{\xi}_{(jj')\eta}$ is resource cost to implement a network section of ξ type between a pair of components (jj') within η period; $M_{(jj')}^{\xi}$ is the duration of the implementation of a transport component of ξ type between a pair of components (jj').

So, the problem is formulated that arises while building the models of topological structures that take into account the dynamics of development of specific components of ShS CN. A mathematical model is proposed to analyse the evolution of the topological structure of ShS core network within a fixed time interval. The proposed model takes into account the dynamics of the development of specific network components, depending on their type and purpose.

6.2 Mathematical Model of the Evolution of ShS CN Architecture

Planning the evolution of ShS core network is a modern important issue that is not formalized sufficiently and is determined, first of all, by dynamically changing needs of subscribers of such systems, technical characteristics of communication channels that are used and existing components [38, 41].

While planning the evolution of ShS CN architecture, the following parameters should be considered:

- a set of components, including sources of information and information addressees, characterized by the rate of information streams and needs, respectively, in the current planning period;
- a set of connections between components, as well as their nature and quantity.

This approach enables representing the evolution process as a certain set of paths on a multipartite alternative graph. The graph vertices, in this case, represent the composition of the technical means available in a particular component within the planning period, and the arcs reflect the capabilities of the corresponding transitions.

Thus, the task of planning the evolution of ShS CN architecture is to find the optimal plan for its development, taking into account the moment of putting new components of the network architecture into operation in each evolution planning period as well as the rate of information streams. Such a task can be represented by a linear task of mathematical programming. For convenience, the following variables are introduced:

- x_{ikt} is a Boolean variable that takes the arithmetic mean value only in the case when i component k composition of technical means within t planning period;
- y_{it} is a rate of the information stream between two components of different levels of hierarchy within t planning period;
- Q_{it} is a rate of the information stream from i component within t period;
- Q_k—productivity of technical means of k-th component;
- I_R is a number of components of the upper level of hierarchy;
- I_j is a number of the next hierarch level that have the communication channels with component j of the highest level of the hierarchy;

- I is a total number of components in the core network of ShS;
- c_{it} is the bandwidth of the communication channel of component i within the period of evolution planning t;
- d_{it} is a factor that reflects the average volume of transmitted information per unit of component performance, and depends on the structure of the tasks the component solves within the period of evolution planning t;
- K_{it}^* is the minimum required set of technical means that are implemented in component i to meet the needs of subscribers within the period of evolution planning t;
- M_t is the maximum number of components of ShS core network of that which is not still put into operation within the period of evolution planning t;
- R_k is the capital cost to implement a component with a set of technical means k;
- R_i is the contribution degree of i component into capital costs to implement the communication channel;
- μ_i is a capital cost increase factor under unfavorable conditions.

Let $q_{ikt} = Q_{it} - Q_k$, and introduced variables are

$$\overline{q}_{ikt} = \begin{cases} q_{ikt}, & \text{if } q_{ikt} > 0; \\ 0 \text{ else,} \end{cases} \qquad \tilde{q}_{ikt} = \begin{cases} -q_{ikt}, & \text{if } q_{ikt} < 0; \\ 0 \text{ else.} \end{cases} \tag{85}$$

Then, the expression for the load balance within the network is:

$$\sum_{t=1}^{T} \sum_{j=1}^{I_R} \sum_{k=1}^{K} \left(\overline{q}_{jkt} x_{jkt} + \sum_{i \in I_j} (\tilde{q}_{ikt} x_{ikt} - y_{it}) \right). \tag{86}$$

The condition to select a certain composition of technical means to implement a component of ShS core network is:

$$\sum_{k=1}^{K} x_{ikt} = 1, i = \overline{1, I}, \ t = \overline{1, T}. \tag{87}$$

A semantic constraint to the rate of information streams between the components of the hierarchy neighbouring levels states that the aggregated information stream from the components of a higher level of the hierarchy to components of a lower level should not exceed the performance of component j of a higher level is:

$$0 \leq \sum_{i \in I_j} y_{it} \leq \sum_{k=1}^{K} \tilde{q}_{jkt} x_{jkt}, j = \overline{1, I_R}, \ t = \overline{1, T}. \tag{88}$$

The next semantic constraint lies in the fact that the productivity increment of component i of the lowest level of the hierarchy should not exceed its productivity deficit, that is

$$0 \leq y_{it} \leq \sum_{k=1}^{K} q_{ikt} x_{ikt}, i = \overline{1, I}, t = \overline{1, T}, \tag{89}$$

under the corresponding constraint on the information stream rate in the channel i, which is associated with the component within the period of evolution planning t:

$$c_{it} \geq d_{it} y_{it}. \tag{90}$$

In its turn, the composition of technical means of component i within the planning period t is constrained by this condition:

$$\sum_{k=1}^{K} k x_{ikt} \leq K_{it}^{*}. \tag{91}$$

Then, the condition can be written for the minimum required number of components in the ShS core network of within the period of evolution planning t:

$$\sum_{i=1}^{I} x_{i1t} \leq M_{t}. \tag{92}$$

Therefore, the expression that describes the sum of all capital costs to implement the ShS core network is:

$$\sum_{t=1}^{T} \sum_{j=1}^{I_R} \left(\mu_j \sum_{k=1}^{K} R_k x_{jkt} - R_k x_{jk(t-1)} \right) + \sum_{i \in I_j} \mu_i \left(\sum_{k=1}^{K} R_k x_{ikt} - R_k x_{ik(t-1)} \right). \tag{93}$$

The optimality criterion, while solving the task for planning the evolution of ShS core network architecture can be represented by:

- expression (86), if the task takes into account the importance of solving the task stages, there are user priorities as well as constraints on costs in the process of evolution;
- expression (93), if the task has load balance constraints over the periods of evolution planning inside the ShS Core network.

Thus, the section deals with probable solutions regarding planning the evolution of the ShS core network (at the architecture level). Approaches to the formalization of tasks related to building the models of architecture evolution are proposed as well as the model of the architecture evolution of ShS core network.

7 Conclusion

The chapter proposes a set of ShS models that enable taking into account the features of Self-healing Systems, in particular, the mechanism of monitoring and troubleshooting, the auto-recovery protocols. The proposed models also enable planning approaches to improve QoS parameters and decrease costs to operate the core network of Self-healing Systems.

The major results are:

1. The general principles of building the structures of Self-healing Systems are studied. The necessity of modelling the information and technical structures of Self-healing Systems is grounded and a diagram of the analysis of the core network structure is given.
2. A mathematical model of the information structure of Self-healing Systems is proposed. A fast algorithm to determine the membership of the current solution to the space of feasible solutions is developed. The developed model takes into account the features of Self-healing Systems and enables taking into account informational relationships between its components. The method of optimization of the developed information structure is further developed. The balance of the load on the network nodes is chosen as a quality criterion, and, accordingly, the sum of deviations of the load of the nodes from the average load—as an indicator. The objective function and constraints of the corresponding optimization task are developed. To solve it, a genetic algorithm is proposed, which enables solving this task much faster for large dimensions.
3. A model of the technical structure of the core network of Self-healing Systems is proposed. A feature of this model is that it is isomorphic to the information model and is built on its basis. To form a technical structure, a stratified three-layer structure is proposed.
4. The peculiarities of information transmission in wireless components of Self-healing Systems are considered. The traffic models of wireless components of Self-healing Systems, focused on the features of such systems, are developed. It is shown that the use of the proposed modifications enables increasing the bandwidth of the TCP in wireless components due to the redistribution of the bandwidth of the core network component.
5. A problem is formulated that arises when building models of ShS topological structures that take into account the dynamics of the development of specific components of the core network. A mathematical model is proposed that takes into account the dynamics of the development of specific network components depending on their type and purpose. Probable solutions are also proposed regarding the planning of the evolution of Self-healing Systems.

References

1. Sidiroglou, S., Laadan, O., Perez, R., Viennot, N., Nieh, J., Keromytis, D.: ASSURE. Automatic software self-healing using REscue points. In: Proceedings of the 14th International Conference on Architectural Support for Programming Languages and Operating Systems, ASPLOS 2009, Washington, DC, USA, March 7–11. **44**(3), 37–48 (2009). https://doi.org/10.1145/2528521. 1508250
2. Frei, R., McWilliam, R., Derrick, B., Purvis, A., Tiwari, A., Serugendo, G.D.M.: Self-healing and self-repairing technologies. Int. J. Adv. Manuf. Technol. **69**, 1033–1061 (2013). https://doi.org/10.1007/s00170-013-5070-2
3. Ardagna, D., Cappiello, C., Fugini, M.G., Mussi, E., Pernici, B., Plebani, P.: Faults and Recovery Actions for SelfHealing Web Services (2006). https://www.academia.edu/20153099/Faults_and_recovery_actions_for_self-healing_web_services
4. Hudaib, A.A., Fakhouri, H.N.: An automated approach for software fault detection and recovery. Commun. Netw. **08**(03), 158–169 (2016). https://doi.org/10.4236/cn.2016.83016
5. Kovalenko, A., Kuchuk, H.: Methods for synthesis of informational and technical structures of critical application object's control system. Adv. Inf. Syst. **2**(1), 22–27 (2018). https://doi.org/10.20998/2522-9052.2018.1.04
6. Ghosh, D., Sharman, R., Rao, H.R. Upadhyaya, S.: Self-healing systems—survey and synthesis. Decis. Support Syst. **42**(4), 2164–2185 (2007). https://doi.org/10.1016/j.dss.2006.06.011
7. Georgiadis, J., Kramer, M.J.: Self-organizing software architectures for distributed systems. In: Proceedings of the 1st Workshop on Self-Healing Systems, Charleston, 18–19 Nov 2002, pp. 33–38 (2002). https://doi.org/10.1145/582128.582135
8. Fakhouri, H.: A survey about self-healing systems (desktop and web application). Commun. Netw. **9**(01), 71–88 (2017). https://doi.org/10.4236/cn.2017.91004
9. Anand, M., Chouhan, K., Ravi, S., Ahmed, S.M.: Context switching semaphore with data security issues using self-healing approach. Int. J. Adv. Comput. Sci. Appl. **2**(6), 55–62 (2011) https://doi.org/10.14569/IJACSA.2011.020608
10. Carzaniga, A., Gorla, A., Pezzè, M.: Self-healing by means of automatic workarounds. In: SEAMS'08, Leipzig, 12–13 May 2008 (2008). https://doi.org/10.1145/1370018.1370023
11. Mukhin, V., Kuchuk, N., Kosenko, N., Kuchuk, H., Kosenko, V.: Decomposition method for synthesizing the computer system architecture. Adv. Intell. Syst. Comput. AISC **938**, 289–300 (2020). https://doi.org/10.1007/978-3-030-16621-2_27
12. Aldrich, J., Sazawal, V., Chambers, C., Nokin, D.: Architecture centric programming for adaptive systems. In: Proceedings of the 1st Workshop on Self-Healing Systems, Charleston, 18–19 Nov 2002, pp. 93–95 (2002). https://doi.org/10.1145/582128.582146
13. Zaitseva, E., Levashenko, V.: Construction of a reliability structure function based on uncertain data. IEEE Trans. Reliab. **65**(4), 1710–1723 (2016)
14. Rabcan, J., Levashenko, V., Zaitseva, E.K., Vassay, M.: Review of methods for EEG signal classification and development of new fuzzy classification-based approach. IEEE Access **8**, 189720–189734 (2020)
15. Zaitseva, E., Levashenko, V.: Multiple-Valued Logic mathematical approaches for multi-state system reliability analysis. J. Appl. Log. **11**(3), 350–362 (2013)
16. Donets, V., Kuchuk, N., Shmatkov, S.: Development of software of e-learning information system synthesis modeling process. Adv. Inf. Syst. **2**(2), 117–121 (2018). https://doi.org/10.20998/2522-9052.2018.2.20
17. Merlac, V., Smatkov, S., Kuchuk, N., Nechausov, A.: Resourses distribution method of university e-learning on the hypercovergent platform. In: 2018 IEEE 9th International Conference on Dependable Systems, Service and Technologies, DESSERT'2018, Kyiv, pp. 136–140 (2018). https://doi.org/10.1109/DESSERT.2018.8409114
18. Kuchuk, G., Kovalenko, A., Komari, I.E., Svyrydov, A., Kharchenko, V.: Improving big data centers energy efficiency. Traffic based model and method. Stud. Syst., Decis. Control **171**, 161–183 (2019). https://doi.org/10.1007/978-3-030-00253-4_8

19. Attar, H., Khosravi, M.R., Igorovich, S.S., Georgievan, K.N., Alhihi, M.: Review and perfor-mance evaluation of FIFO, PQ, CQ, FQ, and WFQ algorithms in multimedia wireless sensor networks. Int. J. Distrib. Sens. Netw. **16**(6), 155014772091323 (2020). https://doi.org/10.1177/1550147720913233

20. Katti, A., Di Fatta, G., Naughton, T., Engelmann, C.: Scalable and fault tolerant failure detection and consensus. In: EuroMPI'15, Bordeaux, 21–23 Sept 2015, pp. 1–9 (2015). https://doi.org/10.1145/2802658.2802660

21. Ehlers, J., van Hoorn, A., Waller, J., Hasselbring, W.: Self-adaptive software system monitoring for performance anomaly localization. In: ICAC'11, Karls-ruhe, 14–18 June 2011 (2011). https://doi.org/10.1145/1998582.1998628

22. Sobchuk, V., Zamrii, I., Olimpiyeva, Y., Laptiev, S.: Functional stability of technological processes based on nonlinear dynamics with the application of neural networks. Adv. Inf. Syst. **5**(2), 49–57 (2021). https://doi.org/10.20998/2522-9052.2021.2.08

23. Sánchez, J., Ben Yahia, I.G., Crespi, N.: POSTER: Self-Healing Mechanisms for Software-Defined Networks (2014). https://arxiv.org/abs/1507.02952

24. Kvassay, M., Levashenko, V., Zaitseva, E.: Analysis of minimal cut and path sets based on direct partial Boolean derivatives. Proc. Inst. Mech. Eng. Part O: J. Risk Reliab. **230**(2), 147–161 (2016)

25. Levashenko, V., Zaitseva, E., Kvassay, M., Deserno, T.M.: Reliability estimation of healthcare systems using Fuzzy Decision Trees. In: Proceedings of the Federated Conference on Computer Science and Information Systems (FedCSIS), Gdansk, Poland, 11–14 Sept 2016, pp. 331–340

26. Rabcan, J., Levashenko, V., Zaitseva, E., Kvassay, M., Subbotin, S.: Non-destructive diagnostic of aircraft engine blades by Fuzzy Decision Tree. Eng. Struct. **197** (2019)

27. Kuchuk, H., Kovalenko, A., Ibrahim, B.F., Ruban, I.: Adaptive compression method for video information. Int. J. Adv. Trends Comput. Sci. Eng. 66–69 (2019). https://doi.org/10.30534/ijatcse/2019/1181.22019

28. Semenov, S., Sira, O., Gavrylenko, S., Kuchuk, N.: Identification of the state of an object under conditions of fuzzy input data. In: East.-Eur. J. Enterp. Technol. **1**(4), 22–30 (2019). https://doi.org/10.15587/1729-4061.2019.157085

29. Kuchuk, N., Shefer O., Cherneva G., Ali, A.F.: Determining the capacity of the self-healing network segment. Adv. Inf. Syst. **5**(2), 114–119 (2021). https://doi.org/10.20998/2522-9052.2021.2.16

30. Svyrydov, A., Kuchuk, H., Tsiapa, O.: Improving efficiently of image recognition process: approach and case study. In: Proceedings of 2018 IEEE 9th International Conference on Dependable Systems, Services and Technologies, DESSERT 2018, pp. 593–597 (2018). https://doi.org/10.1109/DESSERT.2018.8409201

31. Kuchuk, G.A., Akimova, Y.A., Klimenko, L.A.: Method of optimal allocation of relational tables. Eng. Simul. **17**(5), 681–689 (2000)

32. Kovalenko, A., Kuchuk H., Kuchuk N., Kostolny J.: Horizontal scaling method for a hyper-converged network. In: 2021 Int. Conference on Information and Digital Technologies (IDT), Zilina, Slovakia (2021). https://doi.org/10.1109/IDT52577.2021.9497534

33. Ruban, I.V., Martovytskyi, V.O., Kovalenko, A.A., Lukova-Chuiko, N.V.: Identification in informative systems on the basis of users' behaviour. In: Proceedings of the International Conference on Advanced Optoelectronics and Lasers, CAOL, Sept 2019, 9019446, pp. 574–577 (2019). https://doi.org/10.1109/CAOL46282.2019.9019446

34. Kharchenko, V., Kovalenko, A., Andrashov, A., Siora, A.: Cyber security of FPGA-based NPP I&C systems. Challenges and solutions. In: 8th International Topical Meeting on Nuclear Plant Instrumentation, Control, and Human-Machine Interface Technologies 2012, NPIC and HMIT 2012: Enabling the Future of Nuclear Energy, 2012, vol. 2, pp. 1338–1349 (2012)

35. Dustdar, P.S.: A survey on self-healing systems. Approaches Syst. **9**(1), 43–73 (2011). https://doi.org/10.1007/s00607-010-0107-y

36. Abdullah, A., Candrawati, R. Bhakti, M.A.C.: Multi-tiered bio-inspired self-healing architec-tural paradigm for software systems. J. Teknologi Maklumat Multimed. **5**, 1–24 (2009)

37. Gorla, A., Pezzè, M., Wuttke, J.: Achieving cost-effective software reliability through self-healing. Comput. Inform. **29**(1), 93–115 (2010)
38. Michiels, S., Desmet, L., Janssens, N., Mahieu, T., Verbaeten, P.: Self-adapting concurrency. In: The DMonA Architecture. Proceedings of the 1st Workshop on Self-Healing Systems, Charleston, 18–19 Nov 2002, pp. 43–48 (2002). https://doi.org/10.1145/582128.582137
39. Zinchenko, O., Vyshnivskyi, V., Berezovska, Y., Sedlaček, P.: Efficiency of computer networks with SDN in the conditions of incomplete information on reliability. Adv. Inf. Syst. **5**(2), 103–107 (2021). https://doi.org/10.20998/2522-9052.2021.2.14
40. Fuad, M.M., Deb, D., Baek, J.: Self-healing by means of runtime execution profiling. In: Proceedings of 14th International Conference on Computer and Information Technology (ICCIT 2011), Dhaka, 22–24 Dec 2011, pp. 202–207 (2012). https://doi.org/10.1109/ICCITe chn.2011.6164784
41. Shin, M.E.: Self-healing component in robust software architecture for concurrent and distributed systems. Sci. Comput. Programm. **57**(1), 27–44 (2005). https://doi.org/10.1016/j.scico.2004.10.003

Methods to Manage Data in Self-healing Systems

Andriy Kovalenko and Heorhii Kuchuk

Abstract The chapter proposes a set of data management methods in Self-healing Systems. The proposed methods are focused on taking into account the features of Self-healing Systems and allow the improvement of the reduced QoS parameters. A method to calculate the bandwidth of network sections and the required amount of buffer memory for the known network topology and given gravity matrix, which provides the required values of failure probability for Self-healing Systems and ensures a minimum message delivery time. A method of reallocating the computing resource network section of Self-healing System is proposed, which enables increasing the efficiency of using the computing resource of the core Network of Self-healing Systems. Taking into account the peculiarities of data processing in wireless components, methods to manage information transmission routes and information transmission for modification of transport protocols of a wireless component of ShS CN are developed. A method to synthesize the models of complexes of data processing programs in Self-healing Systems is proposed, which is based on the formal-logical apparatus of temporal Petri nets. The method uses trace data obtained while monitoring Self-healing Systems.

Keywords Self-healing systems · Jitter · Temporal Petri net · Core network · Bandwidth

1 Method to Assess Jitter in ShS Core Network

The rate of data transmission, a user of ShS CN can access, is a stochastic process $\xi(t)$ and has a probabilistic description. Due to physical reasons, there always exists a constraint of maximum transmission rate [1]:

A. Kovalenko (✉)
Kharkiv National University of Radio Electronics, 14 Nauki ave., Kharkiv 61166, Ukraine

H. Kuchuk
National Technical University "Kharkiv Polytechnic Institute", 2, Kyrpychova str., Kharkiv 61166, Ukraine
e-mail: kuchuk56@ukr.net

$$\xi_{max} = \max \xi(t). \tag{1}$$

The average transmission rate within a transmission session is equal to

$$m = M[\xi(t)] = \frac{1}{T} \int_0^T \xi(t)dt. \tag{2}$$

Then, the burst strength (or a burst factor) is calculated as

$$B = \frac{\xi_{max}}{m} = \xi_{max} \cdot T / \int_0^T \xi(t)dt \tag{3}$$

In ShS core network, the burst strength is one of the main traffic characteristics that enable assessing the required bandwidth of network sections (NS).

But this characteristic does not enable assessing the rate of a random process execution $\xi(t)$ over time since is not connected to its spectral, and, consequently, correlation features. To take into account a time scale of a random function $\xi(t)$, to describe the rate of data transmission, another traffic characteristic is introduced, this is an average peak time T_P.

From the analysis of random processes, it follows that the outlier duration depends on the level at which the measurement is carried out. The reasonable choice of the value of the level of the peak duration measurement enables determining unambiguously such ShS CN parameters as the required amount of buffer memory as well as to assess the frequency of delay time variations (jitter).

The main theoretical aspects are based on the theory of outliers of random processes.

Figure. 1 shows the implementation of a random process $\xi(t)$ T time long (for example, the duration of a transmission session), where C is a fixed level, τ is an outlier duration, s is an outlier area.

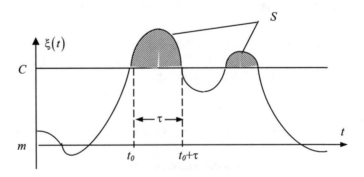

Fig. 1 Example of a random process implementation $\xi(t)$

Actual random processes are continuous functions of time with a limited spectrum due to the inertial properties of both the message source itself and the processing devices acting as a low-pass filter. In particular, the concept of "short and long messages" and the pauses between them are relative and depend on the speed of the switching system implementation. A traffic source is considered bursty if the total time to establish and disconnect the end-to-end channel is less than the interval for the next message. These considerations give grounds to further use the parabolic approximation to derive basic formulas.

Actual random processes are continuous functions of time with a limited spectrum due to the inertial properties of both the message source itself and the processing devices acting as a low-pass filter. In particular, the concept of "short and long messages" and the pauses between them are relative and depend on the speed of the switching system implementation. A traffic source is considered bursty if the total time to establish and disconnect the end-to-end channel is less than the interval for the next message. These considerations give grounds to further use the parabolic approximation to derive basic formulas.

Such functions have a finite number of maxima and minima on a limited time interval T.

The implementation $\xi(t)$ can cross C level upward several times (with positive derivative). The number of such crossings corresponds to the number of positive outliers of a random process above level C, which the theory describe rather strict as a mathematical expression [2]:

$$N^+(C, T) = \int\limits_{o}^{T} dt \int\limits_{o}^{\infty} \xi' \, P(C, \xi'; t) \, d\xi', \qquad (4)$$

where a stroke means differentiation by t.

For stationary random processes, formula (4) is greatly simplified and is

$$N^+(C, T) = P(C) \int\limits_{o}^{\infty} \xi' P(\xi') \, d\xi'. \qquad (5)$$

Formula (5) enables determining the average amount of outliers for the arbitrary distribution law of a random process $\xi(t)$.

If the Gaussian process is stationary, the average value of the total number of crossings by C level process per unit time is

$$N(C) = \frac{1}{\pi}\sqrt{-\rho_o''} \cdot \exp\left[-\frac{1}{2}\left(\frac{c - m}{\sigma}\right)^2\right], \qquad (6)$$

where ρ_o'' is the second derivative of the correlation factor; σ is a standard deviation (SD) of a random process $\xi(t)$.

If levels $C > 0$ are high enough, the majority of outliers have a short duration and are equal to the number of maxima that exceed level C. A random process $\xi(t)$ that is smoothly changing (is differentiated) can be approximated by a parabola in the vicinity of a maximum. Assume that t_o is the moment of a positive outlier start, that is $\xi(t_o) = C$, $\xi'(t_o) > 0$. Let us expand $\xi(t)$ in a Taylor series in the vicinity of t_o point and limit by a quadratic term

$$\xi(t) = \xi(t_o) + \xi'(t_o)(t - t_o) + \frac{1}{2}\xi''(t)(t - t_o)^2. \tag{7}$$

Taking into account that $\xi(t_o) = \xi(t_o - \tau) = C$, the following expression is obtained:

$$\tau = -\frac{2\xi'(t)}{\xi''(t)}. \tag{8}$$

Thus, the task is reduced to calculating the joint probability density $W_o\left(\xi_o', \xi_o''\right)$ of initial derivatives.

Applying the obtained relations to a normal stationary process with correlation function $K(\tau) = \sigma^2 \cdot r(\tau)$, the probability density of outliers durations at high positive levels C is obtained:

$$P(\tau, C) = \frac{1}{4}\left(-\rho_o''\right)\frac{C^2}{\sigma^2}\tau\exp\left[-\frac{1}{8}\left(-R_o''\right)\frac{C^2 \cdot \tau^2}{\sigma^2}\right]. \tag{9}$$

A parabolic approximatanalysis of exion of outliers shape enables calculating the distribution of outliers along the area

$$S = \int\limits_{t_o}^{t_o+\tau} [\xi(t) - C]\,dt. \tag{10}$$

Taking into account expression (7)

$$S = \frac{2}{3}\left(\xi_o'\right)^3\left(\xi''\right)^2. \tag{11}$$

When joint the probability density $W_o\left(\xi_o', \xi_o''\right)$ is known, for the first and second derivatives at the moment of a positive outlier start at level C, the probability density of the distribution of outliers along the area is equal to

$$P(S, C) = \frac{1}{3} \lambda^{2/3} S^{-1/3} \exp\left[-\frac{1}{2}(\lambda S)^{2/3}\right], \tag{12}$$

where,

$$\lambda = \frac{2}{3}\frac{C^2}{\sigma^3}\sqrt{-\rho_o^{//}}, C \gg \sigma. \tag{13}$$

The obtained formula (12) is extremely important since the area of outliers can be considered as the required amount of buffer memory at switching nodes to avoid packet losses and, in this case, level C determines the allowable transmission rate. Packets delayed in memory can be transmitted a the moments when the traffic rate decreases below the set level; moreover, the allowable delay is obviously determined by the requirements of the network temporal transparency, which ensures the service quality for the subscriber who uses the corresponding service.

Let us determine the average value of an outlier area

$$S_{cp} = [MP(C, S)] = \frac{1}{3} \int\limits_{0}^{\infty} S \cdot \lambda^{2/3} S^{-1/3} \exp\left[-\frac{1}{2}(\lambda S)^{2/3}\right] ds = \sqrt{\frac{\pi}{2}} \cdot \frac{3}{\lambda} \tag{14}$$

or, taking into account, (13)

$$S_{cp} = \sqrt{\frac{\pi}{2}} \cdot \frac{9C^2}{2\sigma^3}\sqrt{-\rho_o^{//}}. \tag{15}$$

The average value of an outlier duration, which determines the additional delay due to storing redundant packets in the buffer memory, is

$$\tau_{cp} = \frac{2\psi^2}{\sqrt{-\rho_o^{//} \cdot C}} \tag{16}$$

Let us take the most common correlation coefficient for fractal traffic as an example

$$\rho(\tau) = (1 + \alpha|\tau|) e^{-\alpha(\tau)}, \tag{17}$$

for which $\sqrt{-\rho_0^{//}} = 8\Delta f_э$, where $\Delta f_э$—is the efficient signal bandwidth.

Expressions (6), (9), (12), taking into account (17), are as follows:

$$N(C) = A\Delta f_э \exp\left[-\frac{1}{2}\frac{(C - m)^2}{\sigma}\right], \tag{18}$$

$$P(\tau, C) = 2B\Delta f_{\ni} \cdot C^2 \cdot \tau \exp\left[-B\Delta f_{\ni}\left(\frac{C\tau}{\sigma}\right)\right]; \tag{19}$$

$$\lambda = D \cdot \frac{C^2}{\sigma^3}\Delta f_{\ni}, \tag{20}$$

Moreover, the values of constants A, B, D, depending on a correlation function look, change within the following limits:

$$A = 1.13 \div 2.82$$
$$B = 0.44 \div 1.10$$
$$D = 2.36 \div 5.9.$$

For signal correlation function

$$R(\tau) = \sigma^2 e^{-2\tau} \tag{21}$$

the ultimate expressions of average values of an outlier duration and its area can be presented as:

$$\tau_{cp} = \sqrt{\frac{\pi}{2}} \frac{\sigma}{C \cdot \Delta f_{\ni}}, \tag{22}$$

$$S_{cp} = \frac{1}{2} \cdot \frac{\sigma^3}{\Delta f_{\ni}}. \tag{23}$$

Thus, combining (22) and (23), the average ratio between an outlier duration and its area can be calculated

$$S_{cp} = \frac{1}{\sqrt{2\pi}} \cdot \frac{\sigma^2}{C} \tau_{cp}. \tag{24}$$

If S_{cp} is identified with the required amount of buffer memory $S_{cp} = \overline{S_{6}}$, the average memory amount is expressed through the average peak time as follows

$$\overline{S_{6}} = \frac{1}{\sqrt{2\pi}} \cdot \frac{\sigma^2}{C} \tau_{cp} = \frac{1}{\sqrt{2\pi}} \cdot \frac{\sigma^2}{C} \cdot T_p \tag{25}$$

or, taking into account the standard packet length, the required number of queue positions can be assessed

$$\overline{m_{6}} = 0,75 \cdot 10^{-2} \cdot \frac{\sigma^2}{C} \cdot T_p. \tag{26}$$

However, in practical tasks, it is hardly reasonable to connect the required amount of buffer memory with the average value of an outlier area, since this toughens the requirements for communication channels [3, 4]. The constraint level $C = C_n$ can be determined so that an outlier area with P_o probability does not exceed the set value S_n that is compatible with the available resources of core network of Self-healing Systems. According to expression (12), the probability of an outlier that has an area that exceeds a set value is as follows:

$$P_o(S) = P(S \le S_n) = 1 - \int_{+S_n}^{\infty} P(S, C)\, dS = 1 + e^{-\frac{1}{3}\lambda^{2/3} S_n^{2/3}}, \qquad (27)$$

where λ is calculated by formula (22). Dependence curves (27) when $\lambda = const$ are shown in Fig. 2.

They enable calculating threshold e_n that determines the allowable transmission rate for each service and is given by values ρ_0 and S_n by which parameter λ_n is determined unambiguously:

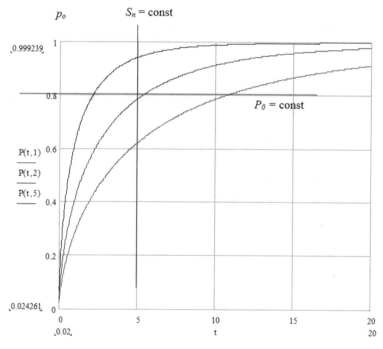

Fig. 2 Dependence curves of an outlier with the area that exceeds a set value ($\lambda = const$)

$$C_n = \sqrt{\frac{\lambda_n \cdot \sigma^3}{D \cdot \Delta t_\vartheta}},$$

(28)

where threshold C_n is counted off from $\xi(t)$, that is $C_n = C - m$, which corresponds to geometrical locus of curve (27) that passes through the point of intersection of lines $P_o = const$ and $S_n = const$.

Thus, the communication is established between traffic spectral features and main parameters of ShS CN. Based on the theory of random processes outliers, the Based on the theory of outliers of random processes, the main calculated ratios are obtained for the classes of traffic which is represented as a stationary normal random process with known average values (dispersion σ^2 and expectation m) and a known correlation function.

The obtained ratios for the average number of outliers per unit time, average values of an outlier duration, and an outlier area enable assessing the allowable jitter range of the core network of Self-healing Systems and, using them, to calculate the bandwidth of network sections.

2   Technique to Calculate the Bandwidth of Sections of the Core Network of Self-healing Systems

While forming the technical structure of the core network of Self-healing Systems, the fundamental problem is the efficient use of network node and channel resources when providing the required quality of servicing users (QoS) [5]. Therefore, the technique to calculate optimal bandwidths of network sections is needed, which takes into account specific features of Self-healing Systems and provides minimum time to deliver data with the given probability of losses.

Each section of ShS CN is modelled as a queueing system (QS) M/M/n with a limited queue, that is a delay-basis n-channel QS, to which a Poisson request stream with total rate λ, servicing rate for each channel μ and a number of queue positions m is fed. Queues are related to the input of each link formed by a bundle of n channels and shared memory in each direction containing m memory buffers.

For each section of ShS CN, the average number of busy channels is determined by the following expression [6]:

$$\bar{z} = \rho \cdot \left(1 - \frac{\rho^{n+m}}{n^m \cdot n!} \cdot P_0\right),$$

(29)

where $\rho = \lambda / \mu$ is reduced rate of the stream of requests (channel load factor).

The average number of requests that are in a queue can be calculated by the following formula [7]:

$$\bar{r} = \frac{(n \cdot \chi)^{n+1} \cdot P_0}{n \cdot n!} \cdot \sum_{\alpha=1}^{m} \alpha \cdot \chi^{\alpha-1}, \tag{30}$$

where $\chi = \rho/n$.

In formulas (29), (30) the value P_0 is determined by the following expression [8]

$$P_0 = \left[\sum_{\alpha=0}^{n} \frac{\rho^\alpha}{\alpha!} + \frac{\rho^n}{n!} \sum_{\alpha=1}^{m} \left(\frac{\rho}{n}\right)^\alpha \right]^{-1} = \left[\sum_{\alpha=0}^{n} \frac{(n \cdot \chi)^\alpha}{\alpha!} + \frac{(n \cdot \chi)^n}{n!} \sum_{\alpha=1}^{m} \chi^\alpha \right]^{-1}. \tag{31}$$

By summing expressions (29) and (30), the average number of requests that are in a QS is obtained:

$$\bar{w} = \bar{z} + \bar{r}. \tag{32}$$

The failure probability in servicing (loss of cells) when all channels and all positions in a queue are occupied is equal to

$$P_{fail} = \frac{\rho^{n+m}}{n! \cdot n^m} \cdot P_0. \tag{33}$$

The failure probability is fixed at an allowable level

$$P_{fail} \leq P_{fail}^{accept}, \tag{34}$$

then, for extreme value P_{fail}^{accept}, P_0 is calculated from expression (33)

$$P_0 = \frac{n! n^m}{(n\chi)^{n+m}} \cdot P_{fail}^{accept}. \tag{35}$$

By using ratio (35), expressions (29) and (30) are simplified:

$$\bar{r} = P_{fail}^{accept} \cdot \sum_{\alpha=1}^{m} \alpha \cdot \chi^{-(m-\alpha)}, \tag{36}$$

$$\bar{z} = n \cdot \chi \cdot \left(1 - P_{fail}^{accept}\right). \tag{37}$$

Finally, using ratio (32), the expression for the average number of requests in QS is obtained

$$\overline{W} = n \cdot \chi \cdot \left(1 - P_{fail}^{accept}\right) + P_{fail}^{accept} \sum_{\alpha=1}^{m} \alpha \cdot \chi^{-(m-\alpha)}. \tag{38}$$

Ratio (38) is fair for any link of an isotropic CN ShS CN, in which the rate of transmitted data ρ does not depend on the transmission direction. However, in anisotropic networks, independent variable χ and values m and n, and, in a general case, even probability P_{fail}^{accept}, depend on the transmission direction for each i-th link, so that

$$\overline{W}_i = n_i \cdot \chi_i \cdot \left(1 - P_{fail}^{accept}\right) + P_{fail}^{accept} \cdot \sum_{\alpha=1}^{m_i} \alpha \cdot \chi_i^{-(m_i - \alpha)}, \quad i = \overline{1, k} \qquad (39)$$

where k is a total number of links in a communication network.

Based on Little formula, according to Kleinrock independence approximation [7], it can be written

$$\gamma \cdot \overline{T}_{giv} = \sum_{i=1}^{k} \overline{W}_i, \qquad (40)$$

where γ is the general traffic in a network; $\overline{T}_{giv}$ is an average packet delay; $\overline{W}_i$ is an average number of packets at the input of each link.

Condition (40), taking into account ratio (39), enables determining such qualitative indicator as the average delay time

$$\overline{T}_{giv} = \frac{1}{\gamma} \cdot \sum_{i=1}^{k} \left[P_{fail}^{accept} \sum_{\alpha=1}^{m_i} \alpha \cdot \chi_i^{-(m_i - \alpha)} + n_i \cdot \chi_i \cdot \left(1 - P_{fail}^{accept}\right) \right]. \qquad (41)$$

The introduction of condition (34) allowed an optimization functional to be significantly simplified due to excluding expression (32), but also introduce another quality indicator—failure probability (losses of packets), the allowable value of which can be introduced as a requirement of network users.

Additionally, function (41) has an extremum (minimum) whose finding is a problem of unconstrained optimization. By calculating partial derivatives

$$\partial \overline{T}_{giv} / \partial \chi_i = 0 \qquad (42)$$

this allows the absolute extremum to be obtained, which, due to function unimodality, is global. In the context of the traditional optimization method, a cost function should be set as a constraint, since the original function does not contain an extremum but is convex, and finding an extremum is solved as a conditional optimization problem with many relative extrema. In addition, this method is free from subjectivity while selecting a cost function, since the use of any of its forms cannot be convincingly argued for specific problem conditions; moreover, using a fibre-optic communication line (FOCL) as a transport medium is a vital necessity independently of its cost and does not have any alternative.

Due to function additivity (41), after calculating partial derivatives (42), a system of algebraic equations is obtained

$$\partial \overline{W}_i / \partial \chi_i = 0, \quad i = 1, k,$$

each of which is the function of one independent variable, that is

$$\partial \overline{W}_i / \partial \chi_i = d \overline{W}_i / d \chi_i = 0. \tag{43}$$

Calculating derivatives (43), taking into account expression (39) results in the system of k equations of the following look:

$$\sum_{\alpha=1}^{m_i-1} (m_i - \alpha) \cdot \alpha \cdot \chi_i^{-(m_i-\alpha+1)} = n_i \cdot \frac{1 - P_{fail}^{accept}}{P_{fail}^{accept}}, \tag{44}$$

each of which enables determining the optimal value $\chi_i = \chi_i^{opt}$ in the function of variables m_i, n_i, P_{fail}^{accept}, which provide the minimum average time of data delivery.

However, under the conditions of the problem, the acceptable values of χ_i^{opt} are only those values that are placed on the surface which is determined by expressions

$$\frac{(n_i \cdot \chi_i)^{n_i+m_i}}{n_i! \cdot n_i^{m_i}} \cdot \left[\sum_{\alpha=0}^{n_i} \frac{(n_i \cdot \chi_i)^{\alpha}}{\alpha!} + \frac{(n_i \cdot \chi_i)^{n_i}}{n_i!} \cdot \sum_{\alpha=1}^{m_i} \chi_i^{\alpha} \right]^{-1} = P_{fail}^{accept}, \tag{45}$$

that were obtained from the constraining condition (33) taking into account (31).

Let us transform expression (44) to the following form:

$$\left[1 + \frac{1}{n_i} \cdot \sum_{\alpha=1}^{m_i-1} (m_i - \alpha) \cdot \alpha \cdot \chi_i^{-(m_i-\alpha+1)} \right]^{-1} = P_{fail}^{accept}. \tag{46}$$

Taking into account that right parts of Eqs. (45) and (46) are the same and constant, acceptable optimal values $^{pr}\chi_i^{opt}$ can be found from conditions when $P_{i\,fail}^{accept} \ll 1$:

$$\frac{n_i!}{n_i \cdot \chi_i} \cdot \sum_{\alpha=0}^{n_i} \frac{(n_i \cdot \chi_i)^{\alpha}}{\alpha!} = \sum_{\alpha=1}^{m_i} \left[\frac{\alpha \cdot (m_i - \alpha)}{n_i} - \chi_i \right] \cdot \chi_i^{\alpha-1}, \quad i = 1, k. \tag{47}$$

The analysis of expression (47) shows that acceptable values $^{pr}\chi_i^{opt}$ do not depend on the required value of the failure probability and are the functions of discrete values of a number of channels (n_i) and a number of positions in a queue (m_i).

Each equation of the system (47) is the function of one variable χ_i and allows the acceptable optimal value $^{pr}\chi_i^{opt}$ to be determined independently for each link of the communication network. However, exact analytical solution of Eq. (47) cannot be worked out due to their transcendence but they can be solved using the special programme by numerical method. In this case, it is sufficient to solve one of the Eq. (47) relatively

$$^{pr}\chi_i^{opt} = \frac{\lambda_i}{\mu_i \cdot n_i} = \frac{L \cdot \lambda_i}{L \cdot \mu_i \cdot n_i} = \frac{F_i}{C_i \cdot n_i} = f(m_i, n_i), \qquad (48)$$

where L is a fixed packet length; $F_i = L \cdot \lambda_i$ is a summing stream at the input of the i-th link; $C_i = L\lambda_i$ is the bandwidth of each of n_i channels.

The rest values $^{pr}\chi_i^{opt}$ can differ just in the number of buffers m_i, or in the number of channels n_i.

The optimization by χ_i allows values C_i and n_i to be varied depending on the class of traffic (F_i), providing a user (on their request) with any channel set with a variable width of bit rates, each time forming a virtual channel with variable bandwidth, regardless of the required failure probability, while data delivery time remains minimal.

In this section, Eq. (47) is solved by a graphical method for isotropic network. The curves of function dependence $F_1(\chi, n)$ and $F_2(\chi, n, m)$, that correspond to the right and left parts of Eq. (19), are presented on the graphs (Fig. 3).

Acceptable optimal values $^{pr}\chi_i^{opt} = Y(m, n)$ are determined by the coordinates of the intersection points of these curves on the χ axis. The analysis of the graphs shows that, depending on the ratio of the values of m and n, three cases are possible.

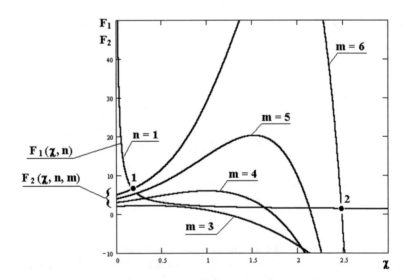

Fig. 3 Curves of functions $F_1(\chi, n)$ and $F_2(\chi, n, m)$

In the first case, curves F_1 (χ, n) and F_2 (χ, n, m) do not cross and the task cannot be solved, that is extremal values of function (41) are not acceptable. In the second case, curves F_1 (χ, n) and F_2 (χ, n, m) have one common point (point of contact), however, similar points for integer n and m are not found.

In the third case, curves F_1 (χ, n) and F_2 (χ, n, m) have two intersection points (1 and 2), however, values $\chi > 1$ at the second point cause the rapid increase of $\overline{T}_{giv}$ and P_{fail}^{accept}. According to the physical meaning and problem condition, values $^{pr}\chi_i^{opt}$, located at point 1, are acceptable.

Thus, values $^{pr}\chi_i^{opt}$, obtained from graphs, along with ratio (48) enable calculating the bandwidths of communication channels and the required amount of buffer memory, with the known network topology and given gravity matrix (λ_{ij}), which provide the required values of the failure probability and ensure the minimum delivery time of messages in ShS CN.

3 Method of Redistribution of the Computing Resource of the Network Section of Self-healing System

3.1 Predicting the Changes of the Data Traffic Rate at the Input of the Network Section of Self-healing System

The aggregated traffic at the input of the network section of Self-healing System depending on the data source can be of different character [9, 10]. However, the fractal traffic characterized by self-similar processes has the greatest impact on the change in QoS parameters.

A factor that is not taken into account in existing methods of computing resource redistribution is the traffic self-similarity, which becomes evident when a lot of information streams combine. Its effect is well described by the factor of deviation of the peak values of the data stream rate that is determined by the expression:

$$k_p = \frac{\lambda_{max}}{\lambda_{average}}, \qquad (49)$$

where λ_{max} is a maximum value of the data stream rate; $\lambda_{average}$ is the average value of the data stream rate.

Theoretical and experimental studies of self-similar processes indicate the possibility of predicting their state in the future, and the closer the value of the Hurst exponent for a given process is to 1, the greater the probability. This is due to the correlation structure that exists in long-term dependent processes. This property can also be used to predict future traffic behaviour on time scales that are appropriate for congestion control. Experimental studies of actual telecommunication traffic routes show a correlation structure on time scales from 250 ms to 20 s, and sometimes

Fig. 4 Data traffic
quantizing and segmentation

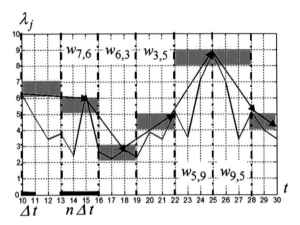

even greater. Taking into account that the time of one transaction is approximately
500 ms (depending on the protocol implementation), the indicated time scales are
adequate to the processes occurring in the network during data exchange, and enable
predicting the future traffic behaviour within this time.

To make predictions, the value of the Hurst exponent is assessed in each stream
of aggregated traffic. If this value does not meet the condition that $0.75 \leq H < 1$,
the known means of the computing resource redistribution are used, the rate values
λ_{pr} for these streams are predicted differently.

The prediction is made in the following way. Within the time interval required to
assess the value of the Hurst exponent, the rate of the j-th stream is quantized by l
levels at the input of λ_j bottleneck section, and the interval is segmented into equal
subintervals with duration $\tau = n\Delta t$, where Δt is the length of a minimum subinterval
of rate change, n is the segment size where the prediction is made (Fig. 4).

For each subinterval with τ duration, the maximum rate value λ_j is chosen.

Let us now form a matrix of frequencies of transitions of the maximum rate values
on each segment from one quantization level onto another

$$W = (w_{\xi,\eta}),$$

where $w_{\xi,\eta}$ is the number of transitions from the ξ-th quantization level onto the η-th
one within the considered time T Let us normalize W row wise [11]:

$$W_{\eta}^{(norm)} = \frac{w_{\xi,\eta}}{\sum_{\xi=1}^{\ell} w_{\xi,\eta}}. \tag{50}$$

In addition, the ξ-th row of matrix $W^{(norm)}$ can be considered as the empirical
discrete function of the probability distribution of transitions from the level/segment
ξ.

Let's consider an example:

$$W = \begin{bmatrix} w_{11} & w_{12} & w_{13} & w_{14} & w_{15} & w_{16} & w_{17} & w_{18} & w_{19} & w_{110} \\ w_{21} & w_{22} & w_{23} & w_{24} & w_{25} & w_{26} & w_{27} & w_{28} & w_{29} & w_{210} \\ w_{31} & w_{32} & w_{33} & w_{34} & w_{35} & w_{36} & w_{37} & w_{38} & w_{39} & w_{310} \\ w_{41} & w_{42} & w_{43} & w_{44} & w_{45} & w_{46} & w_{47} & w_{48} & w_{49} & w_{410} \\ w_{51} & w_{52} & w_{53} & w_{54} & w_{55} & w_{56} & w_{57} & w_{58} & w_{59} & w_{510} \\ w_{61} & w_{62} & w_{63} & w_{64} & w_{65} & w_{66} & w_{67} & w_{68} & w_{69} & w_{610} \\ w_{71} & w_{72} & w_{73} & w_{74} & w_{75} & w_{76} & w_{77} & w_{78} & w_{79} & w_{710} \\ w_{81} & w_{82} & w_{83} & w_{84} & w_{85} & w_{86} & w_{87} & w_{88} & w_{89} & w_{810} \\ w_{91} & w_{92} & w_{93} & w_{94} & w_{95} & w_{96} & w_{97} & w_{98} & w_{99} & w_{910} \\ w_{101} & w_{102} & w_{103} & w_{104} & w_{105} & w_{106} & w_{107} & w_{108} & w_{109} & w_{1010} \end{bmatrix} \quad (51)$$

So, the distribution density of the probabilities of transitions from one quantization level
to another can be considered, according to which the values of the allocated computing resource sufficient for transmitting traffic with the rate λ_j are predicted.

Provided that other transitions in each row, where they are, are also distributed with lower probability, the graphic representation looks like as it is shown in Fig. 5.

The values highlighted in the matrix indicate that the most typical in the considered process are the transitions from the 6-th to the 3-rd, from the 3-rd to the 5th, from the 5th to the 9th and from the 9th to the 5th quantization levels.

This means the greatest probability that after the stream rate at the input of the critical section takes the maximum value that corresponds to the 6-th quantization level, the rate at the next time segment (the interval of prediction) takes a value that does not exceed the 3-rd quantization level. And, as it is shown in Sect. 2, the probability of such a transition is greater, the closer the value of the Hurst exponent for the considered process is to 1.

The combined graph (Fig. 5f) shows the shift of the maximum values of the transition probability.

This shift is characteristic for self-similar processes and characterizes the capability to predict.

Thus, a method is developed to predict the traffic rate at the input of the network section of Self-healing System.

If the value of the rate of incoming flows is known in advance, it is possible to redistribute the computing resource according to predicted.

Therefore, developing a direct method to redistribute computing resources that takes into account the features of the Self-healing System can be considered.

A. Kovalenko and H. Kuchuk

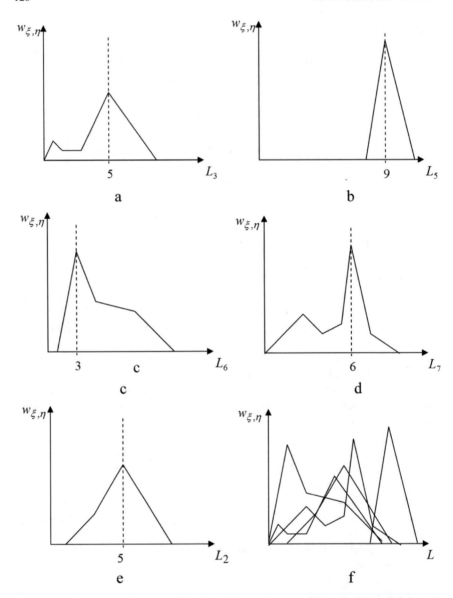

Fig. 5 Densities of transition probability distribution: **a** from the 3rd to the 5th level; **b** from the 5th to the 9th level; **c** from the 6th to the 3rd level; **d** from the 7th to the 6th level; **e** from the 9th to the 5th level; **f** combined graph

3.2 Computing Resource Redistribution

Taking into account the possibility of redistributing the computing resource based on predicting the stream rate at input λ_{pr}, let us consider such a feature as R_{add}—the additional computing resource (CR) obtained within time interval $\tau \ll t$. τ is the interval of time within which the rate value λ_{pr} is predicted, and according to which the bandwidth is redistributed so that the following condition is met

$$\min\left(R_{pr}(\tau)\right) \rightarrow \max\left(\lambda_{pr}(\tau)\right), \tag{52}$$

where R_{pr} is the computing resource redistributed for a data stream with rate λ_{pr} [12].

Proceeding from condition (52), a condition can be written, which is required while redistributing the computing resource taking into account the prediction of the value of data stream (traffic) rate at the input of a critical network section of Self-healing System:

$$\Delta\lambda = \sum_{i=1}^{N_{ch}} \left(\left(P_{allot}^{(i)} + P_{add}^{(i)}\right) - \lambda_{pr}^{(i)}\right) \rightarrow 0, \tag{53}$$

where $R_{allot}^{(i)}$ is an allocated bandwidth for the i-th virtual channel that passes through the network section of Self-healing System $\left(i = \overline{1, N_{ch}}\right)$;

$R_{add}^{(i)}$ is the additional bandwidth for the i-th virtual channel that passes through the network section of Self-healing System;

$\lambda_{pr}^{(i)}$ is a predicted rate for the i-th virtual channel that passes through the network section of Self-healing System.

Let two data streams enter the input of the network section of Self-healing System, these streams have different priorities and rates λ_i and λ_j, and, being combined, make an aggregated stream (Fig. 6) [13].

Then, if data stream with λ_i rate is of higher priority, bandwidth R_i allocated for it corresponds to the maximum value λ_i. Remaining computing power R_j is allocated for data stream with λ_j rate, that is condition $R_{ky} = R_i + R_j$ is met. Besides, $\lambda_{j\ max} > R_j$, which leads to the loss or delay of packets that are quantitively characterized by value v_3 (Fig. 6a).

For the considered case, it is evident that the greater k_n value (49), the smaller computing resource is allocated for data stream with rate λ_j and the more data are lost and they will require retransmitting or will and stay in queue to a data transmission channel. In any case, both leads to an increase in transmission time and a decrease in the efficiency of information exchange at the network section of Self-healing System [14].

Figure 6b illustrates the redistribution of computing resource, which means that for each data stream λ_i, an amount of computing resource is allocated at every time

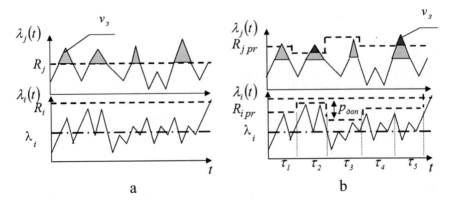

Fig. 6 Graphs that illustrate the principle of the redistribution of the computing resource of the network section of Self-healing System: **a** statically given values R_i, R_j; **b** the redistribution of the computing resource with obtaining R_{add} that is responsible for decreasing the amount of lost or delayed data

interval τ according to (53). Taking into account its high k_n, additional computing resource can be allocated for data stream with rate λ_j:

$$R_{add} = R_i - R_{pr}. \tag{54}$$

Thus, is the computing resource of the network section of Self-healing System R does not change, the section bandwidth increases. The obtained additional computing resource R_{add} can be used either to increase the speed of data stream transmission with rate λ_j or create a virtual route that passes through the network section of Self-healing System. The obtained resource can also be used to transmit the service traffic of ShS CN, whose amount increases when the network structure changes dynamically and can be 3–4 times greater than the summing rate of data streams [15, 16].

After substituting R_{add}—the value of additional CR allocated for a stream with λ_j rate into expression (54), an expression to calculate the packet delivery time at a critical section is obtained taking into account redistributed computing resource

$$T_j = t_{jk} + \frac{V_{j\rho}}{(p_j + p_{add}) \cdot (1 - k_n)} \cdot (n_{jo} + 1). \tag{55}$$

The analysis of the factors that reduce the data exchange rate as an indicator of the efficiency of the task of increasing the data exchange rate in the Self-healing System enables selecting the target function of minimizing the data packet transmission time at a critical section, which is determined by expression

$$\sum_{i=1}^{N_{ch}} T^{(i)} \underset{\rightarrow}{\Omega} \min, \tag{56}$$

where Ω is a set of options for redistributing computing resources at a network section of Self-healing System between virtual channels.

Taking into account (55), the target function to determine the time required to transmit a data traffic packet at a critical section is

$$\sum_{i=1}^{N_{ch}} \left(t_k^{(i)} + \frac{1}{1 - k_n} \cdot \frac{V_{j\rho}}{R_{allot}^{(i)} + R_{add}^{(i)}} \cdot \left(n_o^{(i)} + 1 \right) \right) \to^{\Omega} \min . \tag{57}$$

Moreover, the following constraints should be taken into account:

$$(1) \ \Delta\lambda = \sum_{i=1}^{N_{BK}} \left(\left(R_{allot}^{(i)} + R_{add}^{(i)} \right) - \lambda_{pr}^{(i)} \right) \to^{\Omega} 0; \tag{58}$$

$$(2) \ \sum_{i=1}^{N_{ch}} \left(R_{allot\Omega}^{(i)} + R_{add\Omega}^{(i)} \right) \le k_n \cdot R; \tag{59}$$

$$(3) \ \sum_{i=1}^{N_{BK}} \lambda_{inp}^{(i)} \gg R, \tag{60}$$

where $\lambda_{inp}^{(i)}$ is the data stream rate of the i-th virtual channel.

Expression (57) specifies that an increase in the amount of the allocated computing resource of a network section of Self-healing System, characterized by an increase in bandwidth for the given data stream, results in a decrease in the data packet transmission time due to additional computing resource with a constant bandwidth of a network section.

Thus, a technique to redistribute the computing resource is proposed; it is based on predicting input data streams under limited bandwidth and it allows the features of data traffic self-similarity to be taken into account.

However, this method does not take into account the specificity of wireless components of Self-healing System, which is considered in the following section.

4 Data Processing in Wireless Components of the Core Network of Self-healing Systems

Current numerous studies of various types of traffic and wireless networks of different purposes have shown that it, in most cases, is of a fractal character [17, 18]. In particular, data streams are not of a fractal charter at the beginning, then they are processed at the node servers and active network elements and begin to show clearly defined signs of fractality [19, 20]. One of the fractal traffic features lies in the fact that it has relatively intense bursts at a relatively low average intensity (Fig. 7).

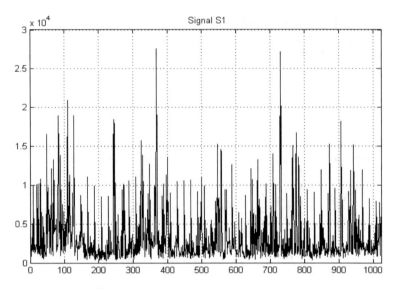

Fig. 7 An example of traffic intensity versus time

Such bursts cause significant delays and packet loss during transmission over the network, even in situations where the total rate of all streams is far from the maximum allowable values. As a result, when using standard network protocols, fractality leads to a significant degradation in both temporal and quantitative characteristics of traffic when it passes through wireless components, since the methods of traffic analysis used nowadays do not provide an adequate picture of processes that occur in the network [21, 22].

To improve the indicators of data streams passage through the network section of Self-healing System (in particular, for wireless components of computer networks (CN) of critical infrastructure entities (CIE), where the fractality effect is the most significant), it is necessary to make changes to the standard traffic control protocols that are based on a short-term traffic prediction.

For a short-term prediction that allows predictable process parameters to be obtained, the fractal traffic features should be taken into account, in particular, the retention of the process structure on a time scale, (self-similarity) [23] and "heavy tails of the distribution" [24] related to a significant effect of autocorrelation (Fig. 8).

For a short-term prediction of the traffic rate of the wireless components of the network section of Self-healing System, which enables taking into account fractal traffic features based on monitoring data streams, it is necessary to have an operational assessment of the Hurst exponent.

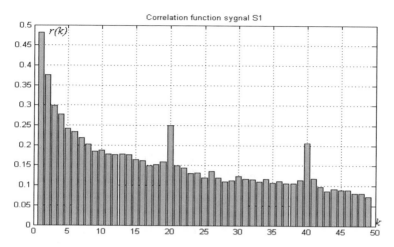

Fig. 8 Traffic autocorrelation (Fig. 7)

4.1 Technique of the Operational Assessment of the Hurst Exponent Based on the Discrete Wavelet Transform

To determine the value of the Hurst exponent, it is proposed to use multi-scale wavelet analysis, the main idea of which is that the considered traffic is decomposed on the orthogonal basis, which is formed by shifts and multi-scale copies of the wavelet function. Discrete wavelet decomposition is splitting the investigated series into two components: approximating and detailing, which are further broken to change the level of decomposition of traffic readings [25]. In the context of multilevel decomposition, the time series is represented by the sum of detailing and approximating components that are determined by the wavelet coefficients of each decomposition level.

Discrete wavelets are usually used along with discrete scaling functions $\varphi_{J,k}(t)$ related to them. Scaling functions have a common task area with wavelets as well as certain relationship between values (shape). According to the discrete wavelet transform, the time series $X(t)$ consists of a set of factors—detailing and low-level approximating ones:

$$
X(t) = approx_J(t) + \sum_{j=1}^{j=J} det\,ail_j(t)
$$

$$
= \sum_k a_{J,k}\varphi_{J,k}(t) + \sum_{j=1}^{J}\sum_k d_{J,k}\psi_{j,k}(t).
$$

(61)

If mother wavelet ψ and corresponding scaling function φ are given, approximating factors $a_{j,k}$ and detailing factors $d_{j,k}$ of the discrete wavelet transform for $X(t)$ process are determined as follows:

$$a_{j,k} = \int\limits_{-\infty}^{\infty} X(t)\varphi_{j,k}(t)dt, \quad d_{j,k} = \int\limits_{-\infty}^{\infty} X(t)\psi_{j,k}(t)dt, \tag{62}$$

where,

$$\varphi_{j,k} = 2^{-j/2}\varphi(2^{-j}t-k); \quad \psi_{j,k} = 2^{-j/2}\psi(2^{-j}t-k). \tag{63}$$

The technique to assess the Hurst exponent is based on the idea that the change in mean values of the squared modules of wavelet factors

$$\mu_j = \frac{1}{n_j}\sum_{k=1}^{n_j}|d_x(j,k)|^2 \tag{64}$$

follows the scaling relationship:

$$\mu_j \sim 2^{(2H-1)j}, \tag{65}$$

where H is the Hurst exponent; $d_x(j, k)$ is detailing wavelet factors at the given level of decomposition j. This ratio allows the Hurst exponent to be assessed:

$$\log_2(\mu_j) = \log_2\left(\frac{1}{n_j}\sum_{k=1}^{n_j}|d_x(j,k)|^2\right) \approx (2H-1)j + const. \tag{66}$$

Thus, if there is a long-term dependence in process $x(t)$, Hurst exponent H can be obtained by estimating the slope of the graph of the function of $\log_2(\mu_j)$ dependence on j. The estimation samples are shown in Fig. 9.

To implement the technique for assessing the Hurst exponent that is based on the discrete wavelet transform, the following sequence of actions is proposed [26].

Step 1. For considered graph X(t), $t = 1, 2, \ldots, n$ the wavelet decomposition is carried out, that is decomposing using wavelet functions with the set maximum level of decomposition.

Step 2. At each level of decomposition j, the factors of detailing $d_x(j, k)$ are determined. The number of detailing factors k are given by the level of decomposition.

Step 3. At each level of decomposition, the sum of squares of wavelet factors is calculated. According to (4.8), the Hurst exponent is in a scaling relationship with the sum of the squares of the modules of the detailing factors.

Step 4. Each level of decomposition j is associated with the logarithm of the sum of squares of the modules of the wavelet factors. Approximating the values of the wavelet factors calculated by the least squares method allows a linear dependence to be obtained, and the Hurst exponent is determined according to its slope.

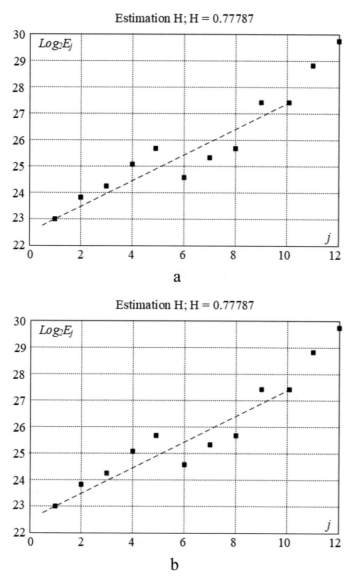

Fig. 9 Assessing the Hurst exponent for the traffic presented in Fig. 7: **a** j = 12, H ≈ 0.78; **b** j = 15, H ≈ 0.80

Thus, the proposed method to determine the value of the Hurst exponent based on the multiple-scale discrete wavelet transform unlike the existing ones, does not require a large amount of computation and allows real-time traffic analysis. Therefore, it is proposed to use it for a short-term prediction of traffic rate in wireless networks.

4.2 Technique to Manage Data Transmission Routes in Wireless Components of the Core Network of Self-healing Systems

The main tasks that are to be solved by the routing methods in wireless components of the Core Network of Self-healing Systems, include [27] selecting routes, distributing traffic along the selected routes, providing proper delivery of messages to addressees. The first two tasks are solved using methods implemented in modern network protocols [28, 29]. The following main characteristics of the wireless components of the Core Network of Self-healing Systems fundamentally affect the correct delivery of messages to addressees [30, 31]:

- bandwidth;
- average delay;
- data transmission time.

The total traffic of CIE CN can be divided into the following categories [27, 32]:

- real-time streams,
- transaction streams;
- data streams.

An increase in the volume of service information in each of these categories inevitably leads to a decrease in the amount of information of Self-healing Systems, especially significantly in the context of an abrupt change in users' data stream rate and the number of failures of the Core Network of Self-healing Systems components.

In addition, when there is such kind of instability, the input buffer memory of routers is congested, including low-priority packets, which results in a sharp increase in packet delays.

One of the widely used routing protocols in up-to-date routers is EIGRP [33]. The EIGRP protocol is an advanced version of the internal routing protocol developed by Cisco to work with the TCP/IP stack protocols and the OSI multi-layer model in data transmission networks, including wireless ones. version of the internal routing protocol developed by Cisco to work with the TCP/IP stack protocols and the OSI multi-layer model in data transmission networks, including wireless ones. The EIGRP protocol is based on a technology that does not require that routers should know all the interconnections in the network, for which a mechanism for declaring addressees using the appropriate distance is implemented. Each router, receiving information, adjusts the distance and transmits information about it to neighbouring routers [34, 35].

Thus, it is crucial to develop a method that enables reducing the data transmission time in wireless components of CIE CN, including during an abrupt change in the rate of traffic streams of various types and with different priorities.

Filling the router buffer memory and the subsequent congestions determine the duration of time intervals in which the load of the router processors will be maximum.

According to the EIGRP protocol, five types of messages are spread in the network [24, 36]:

- Hello: packets to detect "neighbours", that are broadcasted and do not require acknowledgment;
- Update: packets with information about route changes are sent only to routers that are involved in such changes and are confirmed by ACK packets;
- Query: packets to determine the accessibility of the next node on the way to the destination can be sent in any way and are acknowledged by ACK packets;
- Reply: packets that are unicasted in response to Query and acknowledged by ACK packets;
- ACK: packets that acknowledge such packets as Update, Query, Reply.

The EIGRP protocol includes the following components [24, 37]:

- the process of detecting or recovering a neighbouring router that is carried out by a router to recognize adjacent routers in networks dynamically; and along with such a dynamic recognition, routers must recognize the lack of access and failure of adjacent routers using Hello packets;
- a reliable transport protocol that is responsible for guaranteed, ordered delivery of any packet types used in the EIGRP protocol to all adjacent routers and supports both different packet transmission methods and their different priorities;
- the final states machine of the DUAL algorithm that implements the decision-making process while calculating routes by detecting all the routes that are declared by adjacent routers; the selected routes are put in a routing table that is based on the principle of probable further elements (adjacent routers that are used to transmit packets) by the set criterion of optimality of the route to the destination;
- programming modules that are designed to implement network-level functions (for example, interrelating with corresponding protocols, encapsulating EIGRP packets into IP packets, ensuring interoperability with the DUAL protocol, and so on).

Moreover, to implement the EIGRP protocol, the following tables must be supported [24, 38]:

- the neighbour table that is designed for each of the EIGRP protocol modules and used to store information about adjacent routers (as a rule, their addresses, and interfaces), as well as information sufficient to ensure the reliable transport protocol operation;
- the topology table, is used to store and provide information necessary for compiling or modifying the routing table.

All probable routes are divided by the EIGRP protocol into the following groups:

- internal routes: routes that are directly built by the EIGRP protocol;
- external routes: routes that are built by other routing protocols or, static routes that are constantly kept in the routing table.

The proposed technique to manage routes of data transmission in wireless components of CIE CN is based on the EIGRP protocol and aimed at the implementation of rational service data routing; the following sequence of steps is proposed for this.

1. A statistical table is compiled, in which the moments when failures happen are recorded, for example, within 24 h. Data are collected periodically, for example, data collection starts daily at 00:00:00 h.

    The time when the first HELLO message is sent to the network concerning the check of its state t_{fail_1}, is calculated as follows:

    $$t_{fail_1} = \begin{cases} t^{prev}_{fail average}, & t^{prev}_{fail average} < t^{curr}_{fail_1}; \\ t^{curr}_{fail_1}, & t^{prev}_{fail average} \geq t^{curr}_{fail_1}, \end{cases}$$

    where t_{fail_i} is the moments of time when failures happen, $i = \overline{1, n_{fail}}$;
    $t^{prev}_{fail average}$ is the average time interval between failures within the previous 24 h;
    $t^{curr}_{fail_1}$ is the moment of time when the first failure happens within the current 24 h.

2. Based on the accumulated statistical data on failures, the average time interval between failures is calculated $t_{fail_{average}}$.

3. The condition is checked: if the calculated average time interval between failures exceeds the period of 30 min set by the EIGRP protocol to spread an update message on the state of communication channels, at the moment a failure happens, an update message on the state of communication channels is spread instead of failure message:

    - $\ell_{update\_message}$ is spread, if the condition is met:

    $$t_{fail_i} + t_{fail_{average}} > t_{fail_i} + T_{update\_message};$$

    - $l_{failure\_message}$, is spread, if the condition is met:
    - $t_{fail\,i} + t_{fail\,average} \leq t_{fail\,i} + T_{failure\_message}.$

    It should be noted that if the average time interval between failures is over but there was no failure, the message on the state of communication channels is spread nevertheless and changes are recorder to the database, that is if $t_{fail\,average} > T_{update\_message}$, then

    $$t_{spread} = \begin{cases} t_{fail_{i-1}} + t_{fail_{average}}, & t_{fail_{i-1}} + t_{fail_{average}} < t_{fail_i}; \\ t_{fail_i}, & t_{fail_{i-1}} + t_{fail_{average}} \geq t_{fail_i}, \end{cases}$$

    where t_{spread} is the moment of time when the update message on the state of communication channels is spread.

4. It is calculated how often HELLO messages are sent using the average time interval between failures.

The task of determining the failure of network elements of neighbouring routers is assigned to the HELLO message. To calculate the frequency of HELLO messaging, when the average time interval between failures is known, standard queuing systems such as $M/*/*$ and $G_{norm}/*/*$ can be used. M specifies the process where intervals between failures are distributed exponentially, G_{norm} determines the process where intervals between failures are distributed according to the normal law of distribution.

Thus, if the probability of failure is set for the wireless components of the CIE CN as a whole, the frequency of HELLO messaging can be adjusted for interval $(t_{fail_{i-1}}; t_{fail\ i})$, that is while

$$P_{fail}\left(t \in \left(t_{fail_{i-1}}; t_{fail_i}\right)\right) \le P_{fail_{set}},$$

v_{HELLO} can be minimized.

To solve the problem of determining the rational frequency of HELLO messaging, as well as to solve problems linked to assessing the reliability in the system $G_{norm}/*/*$, it is convenient to use the failure probability of network elements at a set interval $(\alpha; \beta)$.

The failure probability of network elements $P_{fail}(t \in (\alpha; \beta))$ at a set interval $(\alpha; \beta)$ is equal to

$$P_{fail}(\alpha < t < \beta) = \int_\alpha^\beta f(t)dt = \frac{1}{\sigma\sqrt{2\pi}} \int_\alpha^\beta \exp^{-\frac{(t-m_t)^2}{2\sigma^2}} dt.$$

The considered systems are basic for wireless networks of various types, but most often the principles of the queuing system $M/*/*$ are used to describe the processes occurring in the network [24, 34]. Note that the method being developed, the accumulation of statistical data on failures is used and, value $t_{fail_{average}}$ is updated constantly.

Consider analytical expressions describing a simplified model $M/*/*$.

At the moment of time $t_{fail_{i-1}}$, let the frequency of failures be close to zero—$P^*(t_{fail_{i-1}}) \cong 0$ (as far as a failure has already occurred), and at the moment of time $t_{fail\ i}$.

Let us consider that the frequency of failures is equal to the frequency of failures at the moment of time $t_{fail_{i-1}} + t_{fail_{average}}$, that is

$$P^*\left(t_{fail_i}\right) \approx P^*\left(t_{fail_{i-1}} + t_{fail_{average}}\right).$$

The frequency of failures at the moment of time t_{fail_i} is equal to

$$P * (t_{fail_i}) \approx P * (t_{fail_{i-1}} + t_{fail_{average}}) = \frac{\frac{T_{observation}}{t_{fail_{average}}} - k_{occur\_fail}}{T_{observation}} \tau, \qquad (67)$$

where $T_{observation}$ is a time interval of statistical data collection, k_{occur_fail} is the number of failures that have occurred at a given time.

If the frequency of failures at both limits of the interval $\left(t_{fail_{i-1}}; t_{fail_i} \right)$ is known and taking into account the properties of the failure stream, the frequency of failures at the elementary time intervals $\tau = 1(\text{sec})$ in the middle of interval $\left(t_{fail_{i-1}}; t_{fail_i} \right)$ can be calculated.

However, based on the accumulated statistical data, it can be concluded that the moment of the next failure is equal to the sum of the time moments of a current failure and the average time interval of a failure, calculated on the basis of statistical data, namely

$$t_{fail_i} = t_{fail_{i-1}} + t_{fail_{average}}. \tag{68}$$

Then, taking into account expressions (67) and (68), the following is obtained:

$$P^*(t_{fail_{i-1}} + s \cdot \tau) = t_{fail_{average}} \cdot s \cdot \tau \cdot \frac{\frac{T_{observation}}{t_{fail_{average}}} - k_{occur\_fail}}{T_{observation}}. \tag{69}$$

According to the method that is used in the EIGRP protocol, the frequency of HELLO messaging is 5 s. It should be noted that after sending It should be noted that after transmitting the current HELLO message and before sending the next message, the router cannot monitor the state of neighbouring routers and, consequently, report its state, that is, it is in a passive state. For EIGRP, the time interval of the router passive state is 15 s.

Proceeding from the above, let time interval between failures $\left(t_{fail_{i-1}}; t_{fail_i} \right)$ be broken into the intervals of 10 s.

The relative frequency that a failure of a network element occurs at the moment when the router is in a passive state is equal to

$$P^*(0 < s \cdot \tau < T_{HELLO}) = s \cdot \tau / T_{HELLO},$$

where T_{HELLO} is the period of spreading the packets of the HELLO message.

The passive state of a router should be considered as a state in which it cannot determine whether its "neighbours" operate or not, that is the router does not send the HELLO message.

The number of such intervals of the passive state at interval $\left(t_{fail_{i-1}}; t_{fail_i} \right)$ is equal to

$$k_{passivity\_interval} = \left(t_{fail_i} - t_{fail_{i-1}} \right) \cdot \nu_{HELLO}$$

Then the average frequency of a network element failure that occurs when the router is in a passive state is

$$P^*_{average}(0 < s\tau < T_{HELLo}) = \frac{\sum_{k=1}^{k_{passivity\_interval}} k_{np} \frac{s \cdot \tau}{T_{HELLO}}}{k_{passivity\_interval}}, \tag{70}$$

where $k_{passivity\_interval}$ is the number of failures that occurred in the passivity interval.

Thus, there is a probability to maximize T_{HELLO}, which allows the amount of service information to be significantly decreased (the amount of HELLO packets information is approximately equal to 70% of all service information spread by a router). Based on the a priori data presented in [24, 34], it is assumed that due to the use of the developed method of decreasing the time of data transmission in wireless components of CIE CN by rational routing the service information, the target function increases approximately by $1.2 - 1.5$, that is, if $l_\rho(l_{message}, l_{frame}) = 2048$ (byte), $\lambda_{nu_{ij}} = 110$ (Mbit/sec), $\lambda_{fail_{ik}} = 46$ (failures per 24 h), it is 48 h for the proposed method and 41 h for the method of the EIGRP protocol.

Thus, the technique to manage data transmission routes is proposed in the section to reduce the data transmission time in wireless components of the Core Network of Self-healing Systems due to the rational service information routing, which enables reducing the data transmission under all constraints as compared to the method used in the EIGRP protocol.

The developed method is adaptive, since it allows the router to regulate the frequency of service information distribution depending on the value of the average interval between failures, calculated on the basis of statistical data.

It should be noted that if the period of statistical data accumulation increases, the accuracy of calculating the average interval between failures increases as well as the efficiency of the proposed method and, accordingly, the data transmission time decreases.

But there is a need to modify the transport protocols of the wireless segment of the Core Network of Self-healing Systems. For this purpose, the method of data transmission is proposed, which is considered in the next section.

4.3 Data Transmission Method to Modify the Transport Protocols of the Wireless Segment of the Core Network of Self-healing Systems

Let us consider the congestion control mechanism of modern transport protocols. Currently, the most widespread methods of information transmission used in the transport layer protocols of wireless networks are methods that assess the bandwidth available for a connection for a wireless communication channel, considering the event of packet loss as the degree of congestion. These methods are implemented, particularly, in the protocols of the TCP Reno family. This approach has a number of identified drawbacks [24, 39], which result in unnecessary packet loss and heavy fluctuations of sliding window size, which negatively affects the use of communication channels, especially wireless ones.

In turn, promising implementations of transport layer protocols, that implement the mechanisms of connection establishment, handshaking and sliding window, which, in particular, include TCP Freeze [24, 40], use fundamentally different methods of information transmission, which control the rate at which packets are sent by the source based on their own assessment of the bandwidth of a wireless communication channel available for connection. To obtain such an assessment, the difference between the expected and actual rates of sending packets to the network is calculated. In this case, a feature that indicates that there is no congestion in the network is the approximate equality between the expected and actual speeds. In case of congestion, the actual speed will be significantly lower than the expected one.

Thus, the implementation of the data transmission mechanism can be reduced to the sequential steps below.

1. Calculating the (theoretically) expected speed at which a source sends packets to the network:

$$V_o = \frac{W(t)}{T_b},$$

   where $W(t)$ is the current size of a sliding window; T_b is the minimum measured value of the packet transmission time for the given connection (as a rule, immediately after the connection is established).

2. Calculating the actual rate at which the source sends packets to the network:

$$V_p = \frac{W(t)}{T_p},$$

   where T_p is the actual (measured) value of the packet transmission time in the established connection.

3. Assessing the number of packets that are in the router queue that the source makes for each acknowledged packet:

$$\delta = (V_o - V_p) \cdot T_b.$$

4. Calculating the next value of a sliding window size based on the speed difference δ, obtained at the previous step, and the value of the sliding window size at the previous time:

$$W(t) = \begin{cases} W(t-1) + 1 & (\delta < \alpha); \\ W(t-1) - 1 & (\delta > \beta); \\ W(t-1) & (\alpha \le \delta \le \beta), \end{cases} \qquad (71)$$

where α is a the smoothing factor in the procedure for calculating the timer cycle time of the retransmission of the data packet; β is the dispersion factor of the packet transmission time in the established connection.

Thus, if the expected and actual source speeds are close, the connection does not use the entire available bandwidth of the wireless channel; to increase the channel utilization factor, it is necessary to increase the speed at which the source sends information. In the case when the actual speed is much lower than expected, the network is overloaded and the speed at which the source sends information must be reduced.

After stabling the connection, the TCP Freeze protocol initiates the slow start phase, and remains in it until the parameter δ reaches a certain statically specified parameter—the threshold value. As long as the condition is met $\delta < \gamma$, there is the exponential γ increase of the sliding window size (for one package per processing cycle).

The next phase is the congestion avoidance phase that is initiated when the size of the sliding window reaches the threshold value, or the condition $\delta > \gamma$ is met.

This phase is characterized by two statically set parameters—the threshold values α and β, and the sliding window size changes according to condition (71).

Data transmission methods can detect a packet loss at any phase if any of the two events occurs:

- the timer for the data packet retransmission is over: the threshold value for leaving the slow start phase is set to half the value of the current sliding window size, and the sliding window size is cleared to a statically set value, then the phase of slow start begins;
- three ACK packets are consecutively received with the same number of the next expected data packet: in this case, the fast retransmission and fast recovery phase is initiated, then the sliding window size is set to a statically specified value and the congestion avoidance phase is initiated.

Along with advantages due to all the innovations (greater efficiency, fewer fluctuations in the sliding window size, and significantly fewer packet retransmissions), there are a number of drawbacks that negatively affect the possibility of the implementation of these innovations in modern wireless networks, which are as follows:

- disproportionate redistribution of the wireless channel bandwidth, with simultaneous traffic streams, based on transmission methods that use the packet loss indicator as a measure of congestion; this disproportion is due to the amount of buffer memory on the route that deals with the packets of a given stream;
- dynamical change in the active connection route, which leads to an unexpected and unpredictable change in the transmission time of the packet, measured by the protocol, and a corresponding change in the size of the sliding window;
- constant congestion in the network due to the fact that the protocol incorrectly estimated the amount of delay during the connection (as a result of the delay in the queuing of packets, which is due to packets from other connections).

To cope with the first drawback, it is possible to use the method [34], which involves using a long sequentially increasing value of the packet transmission time as a signal for rerouting, since such an increase is mainly due to an increase in delay in a new route but not due to congestion. However, finding the optimal values of the protocol parameters, in this case, is a task that cannot be solved. In addition, the proportionality of the channel bandwidth distribution conducted by the data transmission mechanism is also affected by the connection time, since when congestion begins, the value of the measured packet transmission time increases, and the expansion of the sliding window size slows down unlike recently established connections [24].

Consider a new connection that is established after congestion begins on the network. For this connection, value T_b at the $(i + 1)$-th cycle $(T_b(i+1))$ is greater than $T_b(i)$, and, hence, ration $T_b(i + 1)/T_{i+1}$ is greater than ratio $T_b(i)/T_i$, since values T_i i T_{i+1} are almost equal.

The condition to start reducing the sliding window size is [24]

$$W(t) > \frac{\beta}{1 - T_b/T_p}.$$

The condition to start the expansion of the sliding window size is:

$$W(t) < \frac{\alpha}{1 - T_b/T_p}.$$

Let us define the bandwidth of transport protocols connections in wireless networks. The bandwidth of connections in wireless networks is determined by the lowest bandwidth of a channel that belongs to the given route if there is no congestion in the network. Let the queue length in the wireless connection router which had the lowest bandwidth along the set route be $q(t)$ packets. Let the sliding window sizes of traffic sources of TCP Freeze and TCP Reno protocols be equal to $W_v(t)$ and $W_r(t)$ respectively. Then, the least time required to transmit data within the established connection T_b, and actual values of time of packet transmission time $T_R(t)$ can be written as follows:

$$T_b = T_p + \frac{1}{\mu}, \tag{72}$$

$$T_R(t) = T_p + \left(\frac{q(t) + 1}{\mu} \right), \tag{73}$$

where T_p is time for spreading a signal within the connection; μ is a service parameter.

Within the congestion avoidance phase, the sliding window size W_r of the TCP Reno protocol is calculated a follows [41]:

$$W_r = \sqrt{\frac{8}{3p}},\qquad(74)$$

where p is the probability of packets loss.

Expression (74) can be extended as follows:

$$W_r(t) = \frac{K_r}{\sqrt{p(t)}},\quad K_r - constant.\qquad(75)$$

Likewise, the expression for the bandwidth of the TCP Reno protocol, $V_r(t)$, looks like

$$V_r(t) \equiv \frac{W_r(t)}{T_R(t)} = \frac{K_r}{\sqrt{p(t)}} \cdot \frac{1}{T_R(t)}.\qquad(76)$$

Substituting expression (73) into (76), it is obtained:

$$V_r(t) = \frac{K_r}{\sqrt{p(t)}} \cdot \frac{\mu}{\mu \cdot T_p + q(t) + 1}.$$

To analyse the bandwidth of the TCP Freeze protocol at the slow start phase, it is convenient to refer to the number of connection packets that are in the buffers of the intermediate nodes of route δ. Then,

$$\delta = \left(\frac{W_v(t)}{T_b} - \frac{W_v(t)}{T_R(t)}\right).\qquad(77)$$

Taking into account (72) and (73), expression (77) can be rewritten as follows:

$$\delta = W_v(t) \cdot \left(\frac{q(t)}{q(t) + C \cdot T_p + 1}\right) \cdot T_b.$$

Likewise, for the congestion avoidance phase, it is true:

$$W_v(t+1) = \begin{cases} W_v(t) + 1, \ \alpha \cdot \left(\frac{q(t)+\mu T_p+1}{q(t)}\right) > W_v(t); \\ W_v(t), \qquad \alpha \le \delta \le \beta; \\ W_v(t) - 1, \ \beta \cdot \left(\frac{q(t)+\mu T_p+1}{q(t)}\right) < W_v(t), \end{cases}\qquad(78)$$

where α and β are the parameters of the TCP Freeze protocol.

Thus, when the parameters of the TCP Freeze protocol change, the value of product $\alpha \cdot (q(t) + C \cdot T_p + 1)/q(t)$ changes proportionally, which results in the change the sliding window size. Therefore, by changing the values of such statically set parameters, the dynamics of the TCP Freeze protocol can be changed, which has a direct impact on its bandwidth.

One of the conditions for packet loss in wireless networks is the fact that the total amount of traffic exceeds the bandwidth C of the wireless connection for a specific mobile node, or that the traffic decreases dramatically (which may be due to both its movement and the processes taking place in wireless networks). Then, the condition for the lack of losses can be written as:

$$\frac{W_v(t)}{T_R(t)} + \frac{W_r(t)}{T_R(t)} \le C. \tag{79}$$

At the equilibrium point, the following expression can be obtained for $W_v(t)$, according to expression (79):

$$W_v(t) = T_R(t) \cdot C - W_r(t). \tag{80}$$

Substituting expression (75) into (80), it is obtained:

$$W_v(t) = T_R(t) \cdot C - \frac{K_r}{\sqrt{p(t)}}. \tag{81}$$

Replacing expression (72) with (73) and substituting it into (81), it is obtained:

$$W_v(t) = q(t) + T_p \cdot C + 1 - \frac{K_r}{\sqrt{p(t)}}. \tag{82}$$

Thus, expressions (82) and (75) are obtained for the sizes of sliding windows of the TCP Freeze and TCP Reno protocols, respectively, in wireless networks.

Values α and β are constant in the TCP Freeze protocol. The task of the data transmission method at the transport level, implemented by the protocol, is to regulate the sliding window size so that the number of packets in buffers along the connection route lies between the values α and β.

As an improvement, a method is proposed where the values of α and β parameters can change dynamically according to the following principle: while opening connection, the α and β parameters are set by default, then they start to increment or decrement linearly based on the measurements of the packet transmission time in the connection conducted by the protocol. A new variable is introduced to keep the value of the actual instant rate of transmission.

Therefore, the modification of the data transmission method can be reduced to the following steps:

1. Initializing connection and setting the initial values of data transmission parameters.
2. Transmitting information (calculating the expected and actual rates of packet transmission into the network; estimating the number of packets that are in the router queue; calculating the next value of the sliding window size).

3. Analysing an error type while data transmitting.
4. In the case of an error caused by the "handover" event, processing such an event and checking the current value of data transmission..
5. Normalizing data transmission parameters.

- if the value of δ lies within parameters α and β and the packets transmission time for this connection does not increase, parameters α and β are incremented, data transmission goes on and the type of next probable error is checked once (hereinafter—the check of the condition of connection close and transfer to step 2);
- if the value of δ does not lie within parameters α and β and the packets transmission time for this connection increases, parameters α are β are decremented, data transmission goes on and the type of next probable error is checked once (hereinafter—the check of the condition of connection close and transfer to step 2);
- if the value of δ does not lie within parameters α and β and the packets transmission time for this connection does not increase, data transmission goes on and the type of next probable error is checked once (hereinafter—the check of the condition of connection close and transfer to step 2).

The algorithm of the method taking into account the proposed modification is shown in Fig. 10.

Thus, a data transmission method is proposed in the section, which is designed to be used in the protocol of the transport layer TCP Freeze of a wireless segment of the Core Network of Self-healing Systems.

The proposed method enables reducing the data transmission time due to the implementation of the control algorithm of the size of the protocol sliding window which smooths its fluctuations and leads to fewer retransmissions of data packets, and enables processing the "handover" event.

5 Method for Synthesizing the Models of Data Processing Programme Systems in Self-healing Systems

5.1 Technique for Assessing the Performance of Software Complexes

One of the features of Self-healing Systems is the necessity to monitor data processing constantly and rigorously to check whether the system is ready for self-healing. That is why the task to analyse the software quality in the context of assessing its time efficiency is especially important [5, 21]. At the same time, the indicators of Self-healing Systems performance can appear insufficient when solving the required set of tasks under real conditions, which requires changes in both the structure and methods

Fig. 10 Algorithm of data transmission method

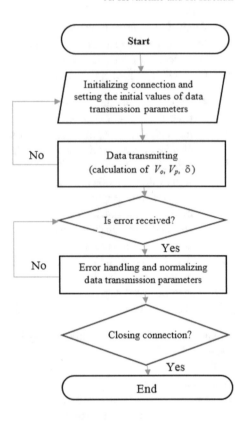

of implementing software, and in the most serious case, hardware. In this regard, the use of various methods for analysing software packages is necessary at all phases of their life cycle that is defined as a set of separate operational stages from developing Self-healing Systems to the termination of its use [16, 22, 27]:

- system analysis and specification of requirements;
- design;
- development (implementation);
- testing;
- maintenance.

As the evidence from practice indicates, in the general case, the phase boundaries of these stages are generally fuzzy, the phases are interdependent, and the process of designing programs is iterative [42, 43], and design changes often occur in parallel to the implementation, which is predetermined by the fact that it is not possible to describe both initial requirements and the system itself completely and in detail [42]. The essential part of the entire life cycle of Self-healing Systems as a software package (Fig. 11) is occupied by verification and validation.

Verification is determined as a process which results in setting the correspondence between the system and established standards (quality indicators) through

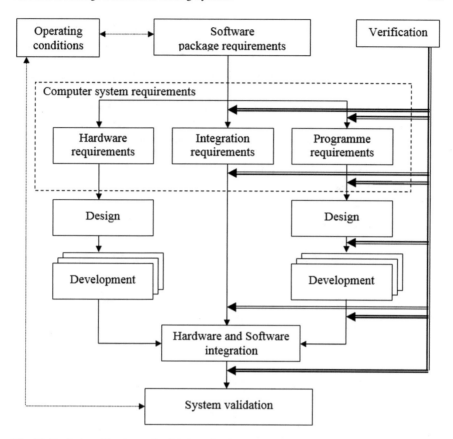

Fig. 11 Design, verification and validation of software packages

formal means. Verification is carried out while passing from one life cycle stage to another one. Validation is defined as the process that determines to what extend the requirements for Self-healing Systems correspond to a running software package when operating conditions are met.

Taking into account the increasing requirements to the quality of software verification, its most objective and complete assessment is needed, which is based on the formation of the system of relations [22]:

- between system requirements and software requirements;
- between Software Requirements and Software functions;
- between Software functions and test methods;
- between tests and verification reports.

At the same time, due to insufficient, incomplete or erroneous specifications of the systems of relations, the software package, even if it has passed the validation and testing, can operate improperly. This is due to the fact that when composing and executing tests, it is necessary to take into account the parallelism of processes that occur in Self-healing Systems, the probability of interrupting the process of calculations and the time when input data arrive as an additional parameter, which makes the task of interpreting test results rather difficult. In addition, the high level of reliability requirements of Self-healing Systems can need unacceptably large amounts of testing.

In [18] formulate the requirement of the parallel (along with the development of the programs) creation of a testing system to boost confidence in the correctness of the system. Building a substantiated mathematical model of software packages and its subsequent analysis at various stages of the life cycle (Fig. 12) must also be a parallel process, allowing timely identification of significant features of the behaviour of the system.

The initial data to model a software package are [22]:

• characteristics of a software package (structural, probabilistic and time);
• characteristics of the physical environment (hardware components of the Core Network of Self-healing Systems and communication systems between them);

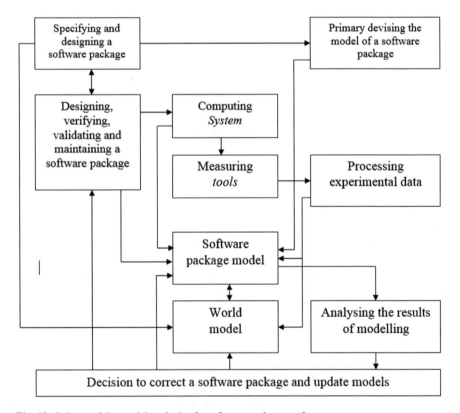

Fig. 12 Scheme of the model analysis of a software package performance

- characteristics of the operating environment as a system for organizing the computing process that provides the execution of user programs (OS, libraries, file system, etc.);
- characteristics of the external environment (parameters of streams of diverse service requests, etc.).

While assessing the performance, statistical data or specified reference values, refined in the process of developing and testing specific programs, are used as specific values of a software package characteristics (Fig. 13).

The major advantage of using a formal model of a software package lies in the fact that its behaviour can be predicted [15]. For example, before the practical implementation of multithreaded programs, preliminary assessments of their time efficiency, the lack of lock-ups and quantitative use of resources (especially processor and memory) can be obtained. The areas critical over processing time can be identified

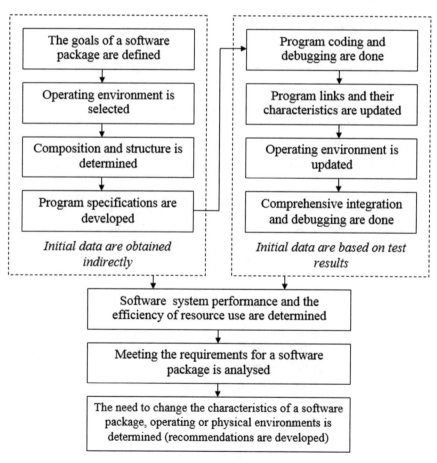

Fig. 13 Stages of analysing the performance of a software package in the process of its development

before they become a bottleneck in the real system, and then, while coding them, the appropriate means can be selected.

The degree of redundancy in systems that should be efficient under failures and errors inside programs is also determined by analysing the model, and both insufficient and excessive redundancy can be found.

Studying the model of a software package at the early stages of design enables:

- determining whether this system meets programmed requirements, as well as if it can be implemented on a certain technical basis;
- monitoring the system integrity;
- identifying subsystems that can run in parallel to achieve optimal performance.

At later stages, structural methods for analysing a model of a software package and its components are used to prove that:

- their implementation is correct;
- all program parts are accessible;
- there are exits from all program parts (there are no lock-ups);
- all busy resources can be released.

The purpose of additional methods of analysis can be to check that:

- the sequence of data processing is correct;
- the results are obtained within a given time interval;
- the values of the program failure probability and losses (for example, time ones) related to their processing are set.

Finally, modelling a software package enables dealing with issues concerning the program adaptivity (capability to be readjusted when external conditions or operating requirements change) and efficient reengineering (capability to be updated) [15].

Thus, providing the required quality of a software package and obtaining the necessary quantitative indicators, including the performance indicator, requires that an adequate formal model and appropriate methods for its subsequent analysis at all stages of the life cycle of a real-time software package be used. Moreover, such a model, due to the iterative nature of the software development, should have an increased degree of flexibility, extensibility and modifiability.

5.2 Synthesizing Models of Software Packages for Data Processing B Self-healing Systems Based on the Generalized Temporal Petri Net

A general approach to the analysis of complex systems is to assess the behavior of its model under the influence of events generated by the model of external influences. In terms of the processes that occur in Self-healing Systems, the set of input tasks (service requests) that are processed according to certain rules is called a workload

[40]. Depending on the objectives of the study, the workload can be understood as both input data and programs (commands). For all known methods of modeling software packages, to describe a workload model that is compatible and corresponding in degree of detail to the system model is rather difficult. In this case, it is necessary to solve an individual task related to the assessment of their accuracy and especially the adequacy [40].

For example, the most widespread graph models of programs are built on the basis of studying the static structures of their initial text but not the actual dynamics of the behaviour of processes, and, in fact, reflect the view of the researcher or developer on the expected system behaviour. Due to this, in the course of modelling, there is a probabity to take into account routes that cannot be actually implemented [42], which reduces the level of confidence in the reliability of the results obtained in the coutrse of modelling. The most representative trace models based on the measured data of actual systems, as mentioned before, are excessively voluminous and not flexible enough.

To eliminate the above disadvantages, a method for the synthesis of network models is proposed, which can be used to create both a workload model and a program complex model [13]. Unlike the known ones, it allows various temporal processes occurring in a computing system to be compactly described on the basis of a single formalization apparatus.

The method is based on the representation of traces of events that occur in the system by temporal Petri nets (TPN). This extension of Petri nets is characterized by using random variables (described by general distribution laws) to determine the delay between transition activating and firing, and by introducing the concept of group transitions that describes a complete group of incompatible events. The probability of a corresponding event corresponds to each transition in the group. The implementation of such a group transition makes it possible to have the process development direction chosen in a non-deterministic way.

The proposed method of the TPN synthesis using trace data comprises five stages.

1. Based on the conceptual model of a software package operation, N types of analyzed temporal events T that are essential in terms of the goals of modelling are singled out:

$$T = \{t_1, t_2, \ldots, tN\}. \tag{83}$$

To build a workload model, the arrivals of dissimilar requests for processing can be considered as events, while in the context of building a software package model—stages of processing requests of specific types according to specified algorithms.

2. Information about the sequence, frequency and execution time of the selected events is collected using a measurement system.

3. Based on the analysis of a received trace, M conditions of a software package are singled out that are determined by the background history:

$$S = \{S_1, S_2..., S_M\}, \ s_i = \left(t_j, \ldots t_k\right) \tag{84}$$

4. The trace is represented by a directed weighted event graph, the vertices T of which are selected states, and the arcs from each vertex reflect the sequence of transitions from one state to another. The weight of each graph arc is defined as the statistical probability of the corresponding transition. It should be noted that such a representation of the event graph corresponds to the Markov chain, which expresses the control structure of the modelled process.
5. Based on the event graph, the TPN is built:

$$NP = \{P, T, F, H, GV, GS, Mo\} \tag{85}$$

where:

- $P = \{p_1, p_2, \ldots, p_L\}$ is a set of positions of the conditions to execute events, that is determined by the input arcs of the event graph;
- T is a set of transitions that correspond to the set of events;
- $F: P \times T \to \{0, 1\}$ is the precedence function of the set of positions P and transitions T that is determined by preconditions to execute events, that is by input arcs of the event graph;
- $H: T \times P \to \{0, 1\}$ is the successor function of the set of positions P and transitions T that is determined by that is by output arcs of the event graph;
- $GV: T \times R_V = f(Z_V)$ is the adherence function between a set of transitions T and a set of stochastic values of duration of events R_V that are distributed by a random law Z_V. The distribution type and parameters are determined based on the trace data using the methods of mathematical statistics;
- $GS: T \times P_S \to [0, 1]$ is the adherence function between a set of group transitions T that form a complete group of incompatible events and a set of probabilities of their firing P_S that are equal to the event graph weights;
- $M_0: P \to \{0, 1, \ldots\}$ is the initial marking;
- $P \cap T = \emptyset$ and $P \cup T \neq \emptyset$.

The weight of each arc is equal to 1, that is the received net ordinary.

An example of the use of the proposed method to model the simplest software package is given below.

Let five groups of particular algorithms be singled out, which process the requests of the same type, to assess the performance of the simplest software package. The aim of the analysis is to receive The goal of the analysis is to obtain the probabilistic-temporal characteristics of the execution of each group of algorithms, which is linked to the corresponding events from the set $T = \{t_1, t_2, \ldots, t_5\}$. Assume that using special measuring tools, time, frequency and the sequence of execution within the specific computing system of programme modules that correspond to particular algorithms are assessed, and the received trace looks as follows:

$$t_2, t_3, t_1, t_2, t_3, t_4, t_2, t_1, t_5, t_2, t_3, t_4.$$

Fig. 14 The view of the
event graph for the analysed
trace

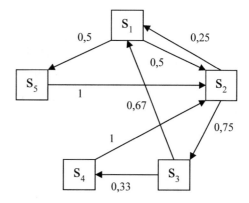

Let us make a simplifying assumption that whether the future event occurs depends only on the current event, that is the given sequence of events can be presented by the Markov chain. This means, that the system conditions are determined by the process history that is one event long, so the set of S conditions of a software package coincides with the set of T.

Based on the trace analysis, an event graph is built (Fig. 14).

As there is no background history, all graph arcs outgoing from the i-th vertex means event t_i, that is why the corresponding designation of events are missing.

The graph arcs outgoing from the vertices s4 and s5 with the probability equal to one mean that the corresponding events are implemented by ordinary time transitions, and the rest—by group transitions. The number of transitions in a group is determined by the number of outgoing arcs, and the probability of their firing is determined by the weight of the corresponding arcs; the sum of the probabilities of group transition events within the construction of the graph is equal to 1.

According to the rules of transformation of an event graph into e Petri net and taking into account a priori information that event t_2 means the start of processing a service request, a net with the initial marking at the vertex P_2 is built (Fig. 15).

In this case, additionally introduced position p_0 that correspond to condition "request has arrived" and transition t_0 that generates requests over random time, act as the simplest workload model and enables analysing the behaviour of a software package under various characteristics of the flow of requests that arrive to be processed.

Without additional position and transition, SDTPN—*Stochastic Deterministic Temporal Petri Net*) is an ordinary active safe Petri net that enables studying closed systems, to which, for example, systems with an established loading regime can be reduced.

The required conditions to use the considered approach to building network models of software packages are:

- the capability to receive accurate information about the sequence and duration of events occurring in the computing system, which supposes firstly the system

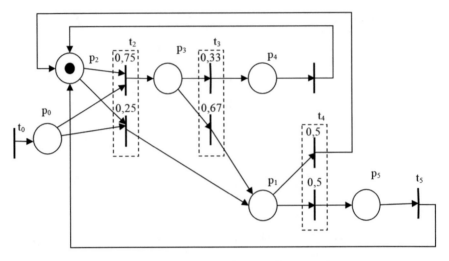

Fig. 15 The temporal Petri net that models the software package operation

algorithms (or at least their main parts) that have been implemented in the program codes, and secondly, special measuring instruments to obtain such assessments, that is measuring monitors [43];

- the assumption that there are no errors and failures while executing programs or that there exist specially developed mechanisms to eliminate their consequences;
- the assumption that the probabilities of system transition by selected states are stationary.

Moreover, to determine the type and parameters of random variables characterizing the processes of program execution accurately, it is necessary to provide an appropriate amount of trace data and certain conditions for their collection, which are considered in [43]. Obviously, to assess the maximum performance of a software package, measurements should be carried out at the maximum load (specified by technical requirements) of the Core Network of Self-healing Systems.

In view of the above, the main drawback the network models synthesized by this method have is their focus on those computing systems on which the measurement information was collected.

The advantages of the proposed method are:

(a) The possibility of using different detail levels while studying various properties of a software package. Since the transitions of the Petri net correspond to the stages of information processing, each of them can be refined by building the corresponding subnets in the course of designing specific programs (and then sub-programs). Thus, the principle of hierarchical modelling is implemented, which makes it possible to increase the understanding of the modelled system and focus on the study of its selected properties;

(b) The use of a single notation to describe both the workload model and the model of the software package itself. The consistency of sub-models (at the level of the formalization apparatus) allows them to be developed separately and provides powerful capabilities to study the behavioural properties of a software package under various conditions and states of the external environment;

(c) The use of a wide variety of types of random variables distributions that characterize the duration of the events selected for analysis, which makes it possible to take into account in the model the main factors affecting the execution process of a software package in a multitasking environment.

The time of programme code execution depends on [22, 26]:

- the amount and values of the initial data;
- difficulties (time complexity) of the computational algorithm and the way of its practical implementation;
- hardware characteristics of the Core Network of Self-healing Systems—the type and structure of the processor, the amount and time of access to RAM, external devices, and so on;
- software characteristics of the Core Network of Self-healing Systems—operating system, file system, programming system, translator from the programming language and so on.

The peculiarities of the computing process in a multitasking OS, for example, in Microsoft Windows, affect this time significantly as well. Pseudo-parallel executable processes constantly compete not only for the processor but also for the RAM, which can lead to the fact that the active program will need additional time to load the necessary part of the executable code or data from the paging file. Time ambiguity can also be caused by queries to standard procedures from the operating system kernel, as well as to other routines stored in dynamically loaded libraries (DLL-files).

Branching, looping, and imperative goto statements can also affect program execution time significantly, depending on whether they are located within the processor pipeline, its first or second level cache, or within conventional memory [9, 18].

Due to the influence of the considered factors, the execution time of a specific program (or a separate part of it), even on the same computer, is a random variable of a very specific type, that differs from exponential or normal.

Thus, the use of general distributions in the TPN allows the peculiarities of programme execution in multitasking computational media to be characterized in the most accurate way, while the proposed method of the network models synthesis enables automating the change data processing and provides:

- a single approach to the formal description of workload and system models, which allows an integrated model to be presented compactly and visually using one of the most powerful modelling tools—temporal Petri nets that operate with general random variables;
- taking into account the background of the development of processes, which enables identifying the interdependencies between events occurring in the computing system;
- the capability to determine not only qualitative (cause-and-effect), but also quantitative (primarily temporal) characteristics of the analysed processes, which makes

it possible to model a variety of real-time software systems, including software packages.

5.3 Method to Assess the Adequacy of the Network Models of Software Packages of Self-healing Systems

One of the main requirements for models is their adequacy to actual systems, which is understood as the ability to predict properly various properties of processes that line up with reality [6]. The processes of operation of actual software packages cannot be described fully and in detail, which is due to their significant complexity. Any model, due to its formality, is adequate only under certain conditions. An increase in the degree of adequacy can be achieved by using various levels of detail, depending on the features of the structural and functional organization of the system and research tasks, taking into account additional factors that affect the process under study, as well as refining the model while designing the system.

The issues of assessing the adequacy of the models and the reliability of the results obtained are detailed in [6, 15]. A method for assessing the adequacy is proposed, which is based on assessing how the model behaviour corresponds to the system, in other words, the adequacy of the description of the dynamics of the process under study.

The model of the software package synthesized by the method proposed above has an important adjustable parameter—M number of system states, determined by the background duration D (the number of transitions taken into account in the system states). As a result of using the hypothesis of the Markov character of the processes under study ($D = 0$), the most compact stochastic model, similar in its capabilities is obtained, and considering the background length of the entire trace leads to the most accurate but cumbersome tracing model.

If the number of analysed events is large, the potential number of system states, the upper boundary of which is defined as $M = C_N^D \cdot D! = N!/(N - D)!$, increases fast if D grows. Thus, when modelling a software package, it is necessary to solve the problem of choosing a specific value of the duration of the considered process background as well as to assess the reliability of the representation of the system behaviour by a trace network model.

To solve this task, as an initial assumption on the nature of the analysed process, let us put forward a hypothesis that this process can be presented by an event graph using a set $S_0 = \{s_0^{(0)}, s_1^{(0)}, \ldots, s_n^{(0)}\}$ of probable system states that coincides with a set T of recorded events ($D = 0$).

The adequacy of the synthesized model should be further analysed and updated according to the following iterative algorithm:

1. Hypothesis "The process of system transitions by states from S_D is Markovian" is checked.
2. The hypothesis is accepted—the algorithm is over, is rejected—go to step 3.

Fig. 16 An example of the behavioural model of a system described by a Markov chain

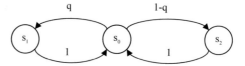

3. D is increased by 1.
4. Set S_D of new system states defined by the process background is determined.
5. The model size is greater than the set limit—the algorithm is over, otherwise—go to step 1.

It should be noted, that despite the lack of aftereffect, the hypothesis about the Markovian behaviour of the system cannot be reduced to the hypothesis about the random nature of sequence of the observed system states, and the statistical criterion of Wald-Wolfowitz series can be used to check these states. For example, to the model of the system shown in Fig. 16, the trace of its behaviour can look like:

$$S_0, S_2, S_0, S_2, S_0, S_1, S_0, S_2, S_0, S_1, S_0, S_1, S_{0,1} \ldots \tag{86}$$

even without any criterion, it is evident that state s_0 is non-random in the trace.

Therefore, it is proposed to test the hypothesis about the Markovian character of the processes of system transition using the theoretical (corresponding to the Markov model) value of the probability that the system reaches a certain selected state in 1, 2, ..., n steps, and the experimental value of the frequency of such events, calculated by the trace data.

Define probability $P_{ij}^{(n)}$ of system transition from the i-th state into the j-th state in n steps ($n \geq 1$) for a network model whose event graph is the Markov chain without absorbing states.

Introduce the generating function

$$\tilde{P}_{ij}(s) = \sum_{n=0}^{\infty} P_{ij}^{(n)} s^n, \quad |s| < 1. \tag{87}$$

If matrix $Q = \|q_{ij}\|$ of transition probabilities of the Markov chain has dimension ρ, the ratio below is obtained by multiplying both parts of the equation by sq_{ki} and summing by $i = \overline{1, \rho}$:

$$s \sum_{i=1}^{\rho} q_{ki} \tilde{P}_{ij}(s) = \sum_{n=0}^{\infty} \sum_{i=1}^{\rho} q_{ki} P_{ij}^{(n)} s^{n+1} = \tilde{P}_{kj}(s) - P_{kj}^{(0)} \tag{88}$$

This ratio determines equation systems

$$\tilde{P}_{kj}(s) - s \sum_{i=1}^{\rho} q_{ki} \tilde{P}_{ij}(s) = P_{kj}^{(0)}, \quad k = \overline{1, \rho}, \quad j = \overline{1, \rho}. \tag{89}$$

Their solutions at fixed k and s are functions of

$$\tilde{P}_{ij}(s) = G_{ij}(s)/D(s)$$

type, which can be decomposed into simple fractions:

$$\tilde{P}_{ij}(s) = \sum_{\lambda=1}^{\rho} \frac{g_{ij}^{(\lambda)}}{1 - s \cdot r_{\lambda}}, \tag{90}$$

where r_{λ} is non-zero characteristic numbers (or eigenvalues) of Q matrix of transition states; $g_{ij}^{(\lambda)}$ is some factors.

Let h be the arbitrary left eigenvector of Q matrix, that is

$$hQ = r_{\lambda}h. \tag{91}$$

Then, the sum of vector elements of the left and right parts of the equation

$$\sum_{j=1}^{\rho} \sum_{i=1}^{\rho} h_i q_{ij} = \sum_{j=1}^{\rho} h_i \sum_{i=1}^{\rho} q_{ij} = \sum_{j=1}^{\rho} h_i = r_{\lambda} \sum_{j=1}^{\rho} h_i. \tag{92}$$

Under the condition that all h_i are positive, $r_{\lambda} = 1$ defines the spectral radius of Q matrix; consequently, all its eigenvalues meet the inequation y $|r_{\lambda}| \le 1$.

Therefore,

$$\frac{1}{1 - s \cdot r_{\lambda}} = \sum_{i=0}^{\infty} s^n r_{\lambda}^n, \tag{93}$$

From (90), it follows

$$P_{ij}^{(n)} = \sum_{\lambda=1}^{\rho} g_{ij}^{(\lambda)} r_{\lambda}^n. \tag{94}$$

To find the values of factors $g_{ij}^{(\lambda)}$, use

$$P_{ij}^{(n+1)} = \sum_{\lambda=1}^{\rho} g_{ij}^{(\lambda)} r_{\lambda}^{n+1} = \sum_{k=1}^{\rho} q_{ik} P_{kj}^{(n)}$$

$$= \sum_{k=1}^{\rho} q_{ik} \sum_{\lambda=1}^{\rho} g_{kj}^{(\lambda)} r_{\lambda}^n = \sum_{\lambda=1}^{\rho} r_{\lambda}^n \sum_{k=1}^{\rho} q_{ik} g_{kj}^{(\lambda)}, \tag{95}$$

from which

$$g_{ij}^{(\lambda)} r_{\lambda} = \sum_{k=1}^{\rho} q_{ik} g_{kj}^{(\lambda)}. \tag{96}$$

On the other hand,

$$P_{ij}^{(n+1)} = \sum_{\lambda=1}^{\rho} g_{ij}^{(\lambda)} r_{\lambda}^{n+1} = \sum_{k=1}^{\rho} P_{ik}^{(n)} q_{kj}$$

$$= \sum_{k=1}^{\rho} \sum_{\lambda=1}^{\rho} g_{ik}^{(\lambda)} r_{\lambda}^n q_{kj} = \sum_{\lambda=1}^{\rho} r_{\lambda}^n \sum_{k=1}^{\rho} g_{ik}^{(\lambda)} q_{kj}, . \tag{97}$$

from which

$$g_{ij}^{(\lambda)} r_{\lambda} = \sum_{k=1}^{\rho} g_{ik}^{(\lambda)} q_{kj}. \tag{98}$$

In matrix form, expressions (96) and (98) can be written as

$$G^{(\lambda)} Q = r_{\lambda} G^{(\lambda)}, \quad Q G^{(\lambda)} = r_{\lambda} G^{(\lambda)}. \tag{99}$$

Thus, matrix columns $G^{(\lambda)}$ are the right eigenvectors of Q matrix and are determined when $r = r_{\lambda}$ by non-zero solutions of $x_i^{(\lambda)}$ equation system

$$\sum_{k=1}^{\rho} q_{ik} x_k - r \cdot x_i = 0; \quad i = \overline{1, \rho}, \tag{100}$$

and the rows are the left eigenvectors that are determined by non-zero solutions of $y_j^{(\lambda)}$ system

$$\sum_{k=1}^{\rho} y_k q_{kj} - r \cdot y_j = 0; \quad j = \overline{1, \rho}. \tag{101}$$

Then, up to a constant factor $C^{(\lambda)}$

$$g_{ij}^{(\lambda)} = C^{(\lambda)} x_i^{(\lambda)} y_j^{(\lambda)}, \tag{102}$$

and immediate value $C^{(\lambda)}$ can be found proceeding from the property of biortonormality of the left and right eigenvectors, according to which

$$C^{(\lambda)} \cdot \sum_{k=1}^{\rho} x_k^{(\lambda)} y_k^{(\lambda)} = 1. \tag{103}$$

Thus, the considered method can be used to determine the theoretical values of probabilities $P_{ij}^{(n)}$ for Markov models.

The frequencies of the system transition from the i-th state into the j-th state in $k = \overline{1, N}$ steps are calculated by the data of the considered trace of the actual system states, they are presented by matrix $V_i = \{v_{ij}^{(k)}\}$, and

$$\sum_{(j)} v_{ij}^{(k)} = n_i^{(k)}. \tag{104}$$

Under the condition that $n_i^{(k)} \geq 50$, as a statistic that characterizes the deviation of experimental frequencies from the corresponding theoretical values, according to Pearson's chi-squared test, the following value can be taken

$$\chi_n^2(V_i) = \sum_{(j)} \frac{[v_{ij}^{(k)} - n_i^{(k)} \cdot P_{ij}^{(k)}]^2}{n_i^{(k)} \cdot P_{ij}^{(k)}} = \sum_{(j)} \frac{[v_{ij}^{(k)}]^2}{n_i^{(k)} \cdot P_{ij}^{(k)}} - n_i^{(k)}. \tag{105}$$

Assigning significance level α, the Markov hypothesis about the behaviour of a system with L number of states is rejected at the exceeded value distributed by the law of chi-squared with L–1 degree of freedom, which corresponds to table value $\chi_{1-\alpha, L-1}^2$.

In addition to Pearson's chi-squared test, which is known to give reliable results only if the row elements of V_j matrix are approximately equal [34] and can be used only when theoretical $P_{ij}^{(k)} \neq 0$ for all k, the information criterion can be used to check if experimental frequencies correspond to theoretical ones at larger L [34]

$$J_C = \frac{\hat{H}_i^{(k)} - M(H_i^{(k)})}{\sqrt{D(H_i^{(k)})}}, \tag{106}$$

where

$$\hat{H}_i^{(k)} = -\sum_{j=1}^{L} \frac{v_{ij}^{(k)}}{n_i^{(k)}} \cdot \ln\left(\frac{v_{ij}^{(k)}}{n_i^{(k)}}\right).$$

is the statistical assessment of the entropy of an empirical distribution;

$$M(H_i^{(k)}) = h_i^{(k)} - (L-1)/n_i^{(k)}.$$

is a mathematical expectation and

$$D(H_i^{(k)}) = \frac{1}{n_i^{(k)}} \left(\sum_{j=1}^{L} P_{ij}^{(k)} \cdot \ln^2(P_{ij}^{(k)}) - [h_i^{(k)}]^2 \right).$$

is the dispersion of the entropy of theoretical distribution;

$$h_i^{(k)} = - \sum_{j=1}^{L} P_{ij}^{(k)} \cdot \ln(P_{ij}^{(k)}).$$

When the empirical frequency distribution coincide with the expected one (except for the uniform distribution law), J_C statistic is a normally distributed with zero expectation and unit variance, and for the given significance level α and the quantile of standardized Gauss distribution $u_{1-\alpha}$, the equation is true

$$|J_C| \le u_{1-\alpha}. \tag{107}$$

Moreover, it should be taken into account that $D(H_i^{(k)}) = 0$ indicates that there is no stochasticity in the system behaviour, which can fully correspond to the selected model.

In terms of power, the information criterion does not actually lie behind the Pearson criterion, while the probability of rejecting a correct hypothesis for some types of distributions is much lower.

The advantage of the considered method is that it can be used not only to synthesze network models of software packages, but also to assess the adequacy of any models described by Markov chains without absorbing states.

The drawback of the considered method is that it is rather difficult to obtain expressions for theoretical probabilities $P_{ij}^{(n)}$, especially if there is a wide variety of system states. In this case, as an alternative to the analytical approach, the values of theoretical probabilities can be obtained by the Monte Carlo method, that is by simulating the behavior of the corresponding Markov chain.

In this case, the problem of testing the hypothesis about the correspondence of experimental and model probability distributions can be solved on the basis of a criterion of the form

$$\chi_n^2(V_i) = n_i^{(k)} m_i^{(k)} \sum_{(j)} \frac{[v_{ij}^{(k)}/n_i^{(k)} - \omega_{ij}^{(k)}/m_i^{(k)}]^2}{v_{ij}^{(k)} + \omega_{ij}^{(k)}}, \tag{108}$$

where $\omega_{ij}^{(k)}$ is the frequencies at which the model passes from the i-th into the j-th state in $k = \overline{1, N}$ steps,

$$\sum_{(j)} \omega_{ij}^{(k)} = m_i^{(k)}. \tag{109}$$

Thus, the developed method, based on the analysis of the processes described by models based on Petri nets, makes it possible to assess their adequacy up to the entered assumptions, to identify cause-and-effect relationships between the system states, to take into account the background of the ongoing processes and to justify the choice of its length.

The use of various criteria when testing hypotheses about the correspondence of theoretical and experimental probabilistic distributions increases the quality of the decision made and reduces the possibility of an irrational increase in the dimension of the synthesized model.

The proposed method can be used to assess the adequacy of a wide class of models, which are based on the description of the process of changing the system states by a Markov chain with absorbing states.

5.4 Aggregate Description of Functional Models of Software Packages of the Self-healing Systems

To determine the conditions and capabilities of using the proposed method to synthesize network models, let us analyse the created TPN in the context of the theory of Petri nets.

The method is based on the distribution of N types of analysed events $T = \{t_0, t_1, \ldots, t_N\}$, which correspond to network transitions. A free language of a Petri net is called L set of all probable sequences of transition firings, that is a set of words composed from the characters of T alphabet events. The trace of events recorded in the computing system using measuring tools is considered as a finite subset $L' \in L$ of words that are generated by a Petri net to be found. Denote the class of any network languages formed by characters of an alphabet defined on all transitions by Λ. It is proved that if $L' \in \Lambda$, the problems of equivalence and selection for such a language cannot be solved. This means that it is not possible to determine whether the language changes, if a certain transition is added into it (or deleted from it). Consequently, for any method on the basis of which the network is built, it is impossible to prove whether it was correct to include or exclude some transitions. Thus, in the general case, it cannot be rigorously proved whether it is correct to represent the event trace with a Petri net.

However, the above problems can be solved for the class of Λ_A—regular languages that are generated, particularly, by finite-state machines. This means that to use the method of synthesizing a TPN, trace data of processes that consist of events that meet the condition of sequence and optionality (but not parallelism) should be analysed. In this case, the unambiguous correspondence between the measured trace and the trace of events generated by the Petri net is justified.

The method of synthesizing a TPN that can be used for sequential-alternative processes, enables solving an inverse problem, which have not been considered previously. It is proposed to define such a transformation as a net-process convolution into a Petri net.

In the context of the structural approach, the most generalized scheme of presenting a complex system is its description as an aggregated system that is formed by combining several aggregates that interact between one another. In our case, an aggregate is understood as a convolution of the i-th process

$$\text{NPi} = \left\{ P^i, T^i, F^i, GV^i, GS^i, M_0^i \right\} \tag{110}$$

where

- zP^i is a set of positions of conditions to execute events of the i-th process;
- T^i is a set of transitions, which corresponds to the set of events of the i-th process;
- $F^i \subseteq P^i \times T^i \cup T^i \times P^i$ is an incidence relation between sets of positions and transitions.
- GV^i is a function of correspondence to transitions of random variables of the execution time of events of the i-th process;
- GS^i is a the function of determining the parameters of group transitions that describe the process developemnt as nondeterministic;
- M_0^i is the initial marking.

For an arbitrary element $a \in A$, define the set of its input elements through $X(a)$, and the set of its output elements through $Y(a)$. Then, structure NP^i meets the condition:

1. $P^i \cap T^i = \emptyset$ and $P^i \cup T^i \neq \emptyset$;
2. $\forall a \in P^i \cup T^i, \exists b \in P^i \cup T^i : b \in X(a) \vee b \in Y(a)$, that is any network element is incident to at least one element of another type;
3. $\forall t \in T^i : X(t) \neq \emptyset \wedge Y(t) \neq \emptyset$, that is any transition has at least one input and one output position;
4. $H^i = \{ p \mid X(p) = \emptyset \}$ is the only position that does not have input transition and that is called a head one;
5. $M_0^i(p) = \begin{cases} 1, & p \in N^i \\ 0, & p \notin N^i \end{cases}$, where N^i is the only position called an initial one.

The TPN of the considered type has all the features of an aggregate [15]: at an arbitrary moment of time, it is in one of the many possible states that are determined by the background and input (control) signals, and produces a finite number of output event signals within a finite amount of time.

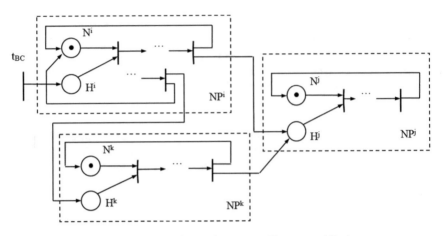

Fig. 17 Aggregate model of processes interaction presented by temporal Petri nets

To introduce relations of succession and parallelism between the processes of such a structure, it is proposed to introduce additional arcs that connect the internal transitions of one network with the head position of another (Fig. 17).

This TPN aggregation is an extension of the concept of action parallel and acyclic networks, or A-nets, that are used to present parallel alternative processes. The appearance of a label at the head position indicates receiving a request to perform a process, which occurs under the condition of its readiness, that is if there is a label at the initial position.

The proposed functional model of the system can be easily augmented with a workload model, which, in this simplest case, is described by transition t_{BC} that generates external service requests.

The obtained aggregated network, in the context of Petri net theory, is finite, correct (that means that it does not have deadlock markings), live (that does not have dead transitions), K-dense and having cyclic components, which is especially distinctive for operating computing systems that are a part of Self-healing Systems.

The considered approach to building the Core Network of Self-healing Systems is based on a conceptual model of the operation of computing facilities as a set of implementations of parallel interacting processes [23, 24]. This allows the behaviour of the system software components to be studied as well as their algorithmic properties.

To expand the modelling capabilities of the aggregate network, namely, to take into account the mechanisms of synchronization of processes and their competition for common resources, additional positions $p_C \in P_C$, can be introduced to the network; they are called the resource ones and they meet the following conditions:

1. $\forall p_C \in P_C$: $M_0^C(P) \geq 1$, that means that all positions-resources have marking that differs from zero one;
2. $\forall p_C \in P_C$: $| X (p_C) | = | Y (p_C) | > 1$, that means that the same number of arcs goes out and enters each resource position, and this number is at least two;

3. $Y(p_C) \in Y(N^i)$, that is output arcs that mean the lock of resources as well as the arcs from the network-process head positions allows the process be executed;

4. $X(p_C) \subseteq T^i$, that is the process while being executed must release the resource.

In the general case, the process assumes several ways of its development, each of which, according to condition 4, must involve releasing the resource, which contradicts condition 2. However, the assumption about the cyclical nature of the processes and the possibility of identifying initial positions determines the possibility to introduce into the TPN structure an additional state that corresponds to the end of the process, and simple (not temporal) transition from this state to the initial one, which does not affect the temporal characteristics of the process.

Determining resource positions allows the competition of processes for the hardware of a computing system to be described in a natural way. For example, the execution of three processes that receive and transmit data, two of which implement the exchange of information with external storage on a hard disk, in a system with two I/O ports can be implemented by an aggregate functional model presented in Fig. 18.

Similarly, the mechanisms of synchronization of processes can be described, which, in modern multitasking systems, are based on the use of mutexes and semaphores—special objects that make it impossible for various processes to access data simultaneously.

It should be noted that unlike the method of synthesizing network models, the introduction of additional positions and arcs leads to the potential danger of lock-ups in the Petri net. This problem can be solved by constructing and analysing a graph of possible network markings.

Thus, the considered methods and conditions for building the models of software packages based on temporal Petri nets, that describe both software and hardware components of computing facilities, make it possible to expand the set of

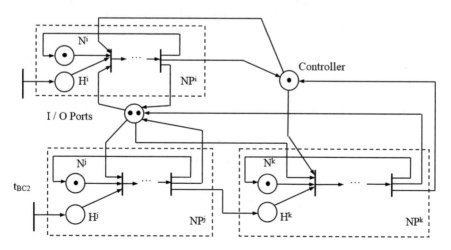

Fig. 18 A version of modelling processes that are competing for resources

analyzed factors that affect the course of the computing process. Confidence in the adequacy of the model is substantiated by the use of measurement traces of executing actual processes that contain information about most of the events occurring in the system. Aggregate description of interacting parallel processes corresponds to modern concepts of design and implementation of software systems in multitasking computing environment.

The concept of processes as minimal independent objects to which resources are allocated, the ways of their presentation and organization are common to most modern computing systems, which provides the generality of the proposed approach to their modelling and the study of various processes in Self-healing Systems.

6 Conclusion

The chapter offers a set of data management methods in Self-healing Systems. The proposed methods are focused on taking into account the peculiarities of Self-healing Systems and allow the reduced QoS parameters to be improved. The major results are the following:

- A method to assess the jitter in the Core Network of Self-healing Systems is proposed. The method is used to calculate the bandwidth of network sections and the required amount of buffer memory for known network topology and a given gravity matrix, which provide the values of the failure probability required for Self-healing Systems and ensure the minimum message delivery time.
- A method to redistribute the computing resource of a network section of the Self-healing System is proposed. The method is based on the proposed method for predicting changes in the information traffic rate at the input of the network section of the Self-healing System and enables increasing the efficiency of using the computing resource of the Core Network of Self-healing Systems.
- The features of data processing in the wireless components of the Core Network of Self-healing Systems are considered. Taking these features into account, a method for the rapid assessment of the value of the Hurst exponent is developed on the basis of a discrete wavelet transform; a method of managing data transmission routes in wireless components, and a method of data transmission to modify transport protocols of a wireless component of the Core Network of Self-healing Systems.
- A method to synthesize the models of data processing programmes in Self-healing Systems is proposed. The method is based on a dynamic model of the behaviour of asynchronous parallel processes and makes it possible to use a single formal logical apparatus—temporal Petri nets—to describe both the system model and the model of the external environment. An increased level of confidence in the synthesized model is achieved when the identified interdependencies between the vents are taken into account; these interdependencies are the system transitions

by the selected states and determined on the basis of trace data obtained in the process of monitoring Self-healing Systems.

References

1. Kuchuk, G., Kovalenko, A., Komari, I.E., Svyrydov, A., Kharchenko, V.: Improving big data centers energy efficiency. Traffic based model and method. Stud. Syst. Decis. Control. **171**, 161–183 (2019). https://doi.org/10.1007/978-3-030-00253-4_8
2. Anand, M., Chouhan, K., Ravi, S., Ahmed, S.M.: Context switching semaphore with data security issues using self-healing approach. Int. J. Adv. Comput. Sci. Appl. **2**(6), 55–62 (2011) https://doi.org/10.14569/IJACSA.2011.020608
3. Michiels, S., Desmet, L., Janssens, N., Mahieu, T., Verbaeten, P.: Self-adapting concurrency. In: The DMonA Architecture. Proceedings of the 1st Workshop on Self-Healing Systems, pp. 43–48. Charleston (2002), 18–19 Nov 2002. https://doi.org/10.1145/582128.582137
4. Fuad, M.M., Deb, D., Baek, J.: Self-healing by means of runtime execution profiling. In: Proceedings of 14th International Conference on Computer and Information Technology (ICCIT 2011), pp. 202–207. Dhaka, 22–24 Dec 2011 (2012). https://doi.org/10.1109/ICCITe chn.2011.6164784
5. Ardagna, D., Cappiello, C., Fugini, M.G., Mussi, E., Pernici, B., Plebani, P.: Faults and recovery actions for self-healing web services (2006). https://www.academia.edu/20153099/ Faults_and_recovery_actions_for_self-healing_web_services
6. Georgiadis, J., Kramer, M.J.: Self-organizing software architectures for distributed systems. In: Proceedings of the 1st Workshop on Self-Healing Systems, pp. 33–38. Charleston (2002), 18–19 Nov 2002. https://doi.org/10.1145/582128.582135
7. Carzaniga, A., Gorla, A., Pezzè, M.: Self-healing by means of automatic workarounds. In: SEAMS'08. Leipzig (2008), 12–13 May 2008. https://doi.org/10.1145/1370018.1370023
8. Ghosh, D., Sharman, R., Rao, H.R. Upadhyaya, S.: Self-healing systems—survey and synthesis. Decis. Support. Syst. **42**(4), 2164–2185 (2007). https://doi.org/10.1016/j.dss.2006.06.011
9. Sánchez, J., Ben Yahia, I.G., Crespi, N.: POSTER: Self-healing mechanisms for software-defined networks (2014). https://arxiv.org/abs/1507.02952
10. Ehlers, J., van Hoorn, A., Waller, J., Hasselbring, W.: Self-adaptive software system monitoring for performance anomaly localization. In: ICAC'11, Karls-ruhe, 14–18 June 2011 (2011). https://doi.org/10.1145/1998582.1998628
11. Svyrydov, A., Kuchuk, H., Tsiapa, O.: Improving efficienty of image recognition process: approach and case study. In: Proceedings of 2018 IEEE 9th International Conference on Dependable Systems, Services and Technologies, DESSERT 2018. pp. 593–597 (2018). https:// doi.org/10.1109/DESSERT.2018.8409201
12. Kuchuk, H., Kovalenko, A., Ibrahim, B.F., Ruban, I.: Adaptive compression method for video information. Int. J. Adv. Trends Comput. Sci. Eng. 66–69 (2019). https://doi.org/10.30534/ija tcse/2019/1181.22019
13. Katti, A., Di Fatta, G., Naughton, T., Engelmann, C.: Scalable and fault tolerant failure detection and consensus. In: EuroMPI'15, pp 1–9. Bordeaux (2015), 21–23 Sept 2015. https://doi.org/ 10.1145/2802658.2802660
14. Aldrich, J., Sazawal, V., Chambers, C., Nokin, D.: Architecture centric programming for adaptive systems. In: Proceedings of the 1st Workshop on Self-Healing Systems, pp. 93–95, Charleston (2002), 18–19 Nov 2002. https://doi.org/10.1145/582128.582146
15. Jiang, M., Zhang, J., Raymer, D., Strassner, J.: A modeling framework for self-healing software systems (2007). https://st.inf.tu-dresden.de/MRT07/papers/MRT07_Jiangl_etall.pdf
16. Kuchuk, N., Shefer, O., Cherneva, G., Alnaeri, F.A.: Determining the capacity of the self-healing network segment. Adv. Inf. Syst. **5**(2), 114–119 (2021). https://doi.org/10.20998/2522-9052.2021.2.16

17. Mukhin, V., Kuchuk, N., Kosenko, N., Kuchuk, H., Kosenko, V.: Decomposition method for synthesizing the computer system architecture. Adv. Intell. Syst. Comput. AISC. **938**, 289–300 (2020). https://doi.org/10.1007/978-3-030-16621-2_27
18. Tkachov, V., Kovalenko, A., Kuchuk, H., Ia, N.: Method of ensuring the survivability of highly mobile computer networks. Adv. Inf. Syst. **5**(2), 159–165 (2021). https://doi.org/10.20998/2522-9052.2021.2.24
19. Pliushch, O., Vyshnivskyi, V., Berezovska, Y.: Robust telecommunication channel with parameters changing on a frame-by-frame basis. Adv. Inf. Syst. **4**(3), 62–69 (2020). https://doi.org/10.20998/2522-9052.2020.3.07
20. Fakhouri, H.: A survey about self-healing systems (Desktop and Web Application). Commun. Netw. **9**(01), 71–88 (2017). https://doi.org/10.4236/cn.2017.91004
21. Sidiroglou, S., Laadan, O., Perez, R., Viennot, N., Nieh, J., Keromytis, D.: ASSURE. Automatic software self-healing using rescue points. In: Proceedings of the 14th International Conference on Architectural Support for Programming Languages and Operating Systems, ASPLOS 2009, vol. 44(3), pp. 37–48. Washington, DC, USA (2009), 7–11 Mar 2009. https://doi.org/10.1145/2528521.1508250
22. Frei, R., McWilliam, R., Derrick, B., Purvis, A., Tiwari, A., Serugendo, G.D.M.: Self-healing and self-repairing technologies. Int. J. Adv. Manuf. Technol. **69**, 1033–1061 (2013). https://doi.org/10.1007/s00170-013-5070-2
23. Merlac, V., Smatkov, S., Kuchuk, N., Nechausov, A.: Resourses distribution method of University e-learning on the hypercovergent platform. In: 2018 IEEE 9th International Conference on Dependable Systems, Service and Technologies, DESSERT'2018, pp. 136–140. Kyiv (2018). https://doi.org/10.1109/DESSERT.2018.8409114
24. Attar, H., Khosravi, M.R., Igorovich, S.S., Georgievan, K.N.: Alhihi, M.: Review and performance evaluation of FIFO, PQ, CQ, FQ, and WFQ algorithms in multimedia wireless sensor networks. Int. J. Distrib. Sens. Netw. **16**(6), 155014772091323 (2020). https://doi.org/10.1177/1550147720913233
25. Momit, O., Zhyvotovskyi, R., Onbinskyi, Y., Lyashenko, A.: Analysis of the known methods of channels communication control with the interference and selective fading. Adv. Inf. Syst. **3**(4), 45–51 (2019). https://doi.org/10.20998/2522-9052.2019.4.06
26. Gorla, A., Pezzè, M., Wuttke, J.: Achieving cost-effective software reliability through self-healing. Comput. Inf. **29**(1), 93–115 (2010)
27. Hudaib, A.A., Fakhouri, H.N.: An automated approach for software fault detection and recovery. Commun. Netw. **08**(03), 158–169 (2016). https://doi.org/10.4236/cn.2016.83016
28. Zaitseva, E., Levashenko, V.: Multiple-Valued Logic mathematical approaches for multi-state system reliability analysis. J. Appl. Log. **11**(3), 350–362 (2013)
29. Sedlacek, P., Zaitseva, E., Levashenko, V., Kvassay, M.: Critical state of non-coherent multi-state system. Reliab. Eng. Syst. Saf. **215** (2021)
30. Kovalenko, A., Kuchuk, H., Kuchuk, N., Kostolny, J.: Horizontal scaling method for a hyper-converged network. In: 2021 International Conference on Information and Digital Technologies (IDT). Zilina, Slovakia (2021). https://doi.org/10.1109/IDT52577.2021.9497534
31. Zaitseva, E., Levashenko, V., Sedlacek, P., Kvassay, M., Rabcan, J.: Logical differential calculus for calculation of Birnbaum importance of non-coherent system, Reliab. Eng. Syst. Saf. **215** (2021)
32. Levashenko, V., Lukyanchuk, I., Zaitseva, E., Kvassay, M., Rabcan, J., Rusnak, P.: Development of programmable logic array for multiple-valued logic functions. In: IEEE Transactions on Computer-Aided Design of Integrated Circuits and Systems, vol. 39(12), pp. 4854–4866 (2020)
33. Attar, H., Khosravi, M.R., Igorovich, S.S., Georgievan, K.N., Alhihi, M.: Review and performance evaluation of FIFO, PQ, CQ, FQ, and WFQ algorithms in multimedia wireless sensor networks. Int. J. Distrib. Sens. Netw. **16**(6), (2020). https://doi.org/10.1177/1550147720913233
34. Aleksandrov, Y., Aleksandrova, T., Kostianyk, I.: Parametric synthesis of the digital invariant stabilizer for a non-stationary object. Adv. Inf. Syst. **4**(1), 39–44 (2020). https://doi.org/10.20998/2522-9052.2020.1.07

35. Zaitseva, E., Levashenko, V., Lukyanchuk, I., Rabcan, J., Kvassay, M., Rusnak, P.: Application of generalized reed–muller expression for development of non-binary circuits. Electronics (Switzerland) **9**(1), (2020), Article number 12, (4)

36. Rabcan, J., Levashenko, V., Zaitseva, E., Kvassay, M.: Review of methods for EEG signal classification and development of new fuzzy classification-based approach. IEEE Access **8**, 189720–189734 (2020)

37. Kuchuk, G.A., Akimova, Y.A., Klimenko, L.A.: Method of optimal allocation of relational tables. Eng. Simul. **17**(5), 681–689 (2000)

38. Rabcan, J., Levashenko, V., Zaitseva, E., Kvassay, M.: EEG signal classification based on fuzzy classifiers. IEEE Trans. Ind. Inf. **18**(2), 757-766 (2022)

39. Semenov, S., Sira, O., Gavrylenko, S., Kuchuk, N.: Identification of the state of an object under conditions of fuzzy input data. East.-Eur. J. Enterp. Technol. **1**(4), 22–30 (2019). https://doi.org/10.15587/1729-4061.2019.157085

40. Dustdar, P.S.: A survey on self-healing systems. Approaches Syst. **9**(1), 43–73 (2011). https://doi.org/10.1007/s00607-010-0107-y

41. Abdullah, A., Candrawati, R. Bhakti, M.A.C.: Multi-tiered bio-inspired self-healing architectural paradigm for software systems. J. Teknologi Maklumat Multimedia **5**, 1–24 (2009)

42. Shin, M.E.: Self-healing component in robust software architecture for concurrent and distributed systems. Sci. Comput. Prog. **57**(1), 27–44 (2005). https://doi.org/10.1016/j.scico.2004.10.003

43. Zhou, J., Wunderlich, H.-J.: Software-based self-test of processors under power constraints. In: Proceedings of Design, Automation and Test in Europe, vol. 1, pp. 1–6. Munich (2006), 6–10 Mar 2006. https://doi.org/10.1109/DATE.2006.243798

Printed in the United States
by Baker & Taylor Publisher Services